C0-AVT-105

# LINKAGES

## MANUFACTURING TRENDS IN ELECTRONIC INTERCONNECTION TECHNOLOGY

Committee on Manufacturing Trends in Printed Circuit Technology
Board on Manufacturing and Engineering Design
Division on Engineering and Physical Sciences

NATIONAL RESEARCH COUNCIL
*OF THE NATIONAL ACADEMIES*

THE NATIONAL ACADEMIES PRESS
Washington, D.C.
**www.nap.edu**

**THE NATIONAL ACADEMIES PRESS**     **500 Fifth Street, N.W.**     **Washington, DC 20001**

NOTICE: The project that is the subject of this report was approved by the Governing Board of the National Research Council, whose members are drawn from the councils of the National Academy of Sciences, the National Academy of Engineering, and the Institute of Medicine. The members of the committee responsible for the report were chosen for their special competences and with regard for appropriate balance.

This study was supported by Contract No. N00014-00-G-0230 between the National Academy of Sciences and the Department of Defense. Any opinions, findings, conclusions, or recommendations expressed in this publication are those of the authors and do not necessarily reflect the views of the organizations or agencies that provided support for the project.

International Standard Book Number 0-309-10034-8

Available in limited quantities from the Board on Manufacturing and Engineering Design, 500 Fifth Street, N.W., Washington, DC 20001, bmed@nas.edu, http://www.nationalacademies.edu/bmed.

Additional copies of this report are available from the National Academies Press, 500 Fifth Street, N.W., Lockbox 285, Washington, DC 20055; (800) 624-6242 or (202) 334-3313 (in the Washington metropolitan area); Internet, http://www.nap.edu.

Copyright 2005 by the National Academy of Sciences. All rights reserved.

Printed in the United States of America

# THE NATIONAL ACADEMIES
*Advisers to the Nation on Science, Engineering, and Medicine*

The **National Academy of Sciences** is a private, nonprofit, self-perpetuating society of distinguished scholars engaged in scientific and engineering research, dedicated to the furtherance of science and technology and to their use for the general welfare. Upon the authority of the charter granted to it by the Congress in 1863, the Academy has a mandate that requires it to advise the federal government on scientific and technical matters. Dr. Ralph J. Cicerone is president of the National Academy of Sciences.

The **National Academy of Engineering** was established in 1964, under the charter of the National Academy of Sciences, as a parallel organization of outstanding engineers. It is autonomous in its administration and in the selection of its members, sharing with the National Academy of Sciences the responsibility for advising the federal government. The National Academy of Engineering also sponsors engineering programs aimed at meeting national needs, encourages education and research, and recognizes the superior achievements of engineers. Dr. Wm. A. Wulf is president of the National Academy of Engineering.

The **Institute of Medicine** was established in 1970 by the National Academy of Sciences to secure the services of eminent members of appropriate professions in the examination of policy matters pertaining to the health of the public. The Institute acts under the responsibility given to the National Academy of Sciences by its congressional charter to be an adviser to the federal government and, upon its own initiative, to identify issues of medical care, research, and education. Dr. Harvey V. Fineberg is president of the Institute of Medicine.

The **National Research Council** was organized by the National Academy of Sciences in 1916 to associate the broad community of science and technology with the Academy's purposes of furthering knowledge and advising the federal government. Functioning in accordance with general policies determined by the Academy, the Council has become the principal operating agency of both the National Academy of Sciences and the National Academy of Engineering in providing services to the government, the public, and the scientific and engineering communities. The Council is administered jointly by both Academies and the Institute of Medicine. Dr. Ralph J. Cicerone and Dr. Wm. A. Wulf are chair and vice chair, respectively, of the National Research Council.

**www.national-academies.org**

## COMMITTEE ON MANUFACTURING TRENDS IN PRINTED CIRCUIT TECHNOLOGY

DAVID J. BERTEAU, *Chair,* Clark and Weinstock
KATHARINE G. FRASE, IBM Microelectronics
CHARLES R. HENRY, U.S. Department of Defense (retired)
JOSEPH LaDOU, University of California, San Francisco
KATHY NARGI-TOTH, Technic, Inc.
ANGELO M. NINIVAGGI, JR., Plexus Corporation
MICHAEL G. PECHT, University of Maryland
E. JENNINGS TAYLOR, Faraday Technology, Inc.
RICHARD H. VAN ATTA, Institute for Defense Analyses
ALFONSO VELOSA III, Gartner, Inc.
DENNIS F. WILKIE, Compass Group, Ltd.

**Staff**

TONI MARECHAUX, Study Director
MARTA VORNBROCK, Research Assistant
LAURA TOTH, Senior Program Assistant

**BOARD ON MANUFACTURING AND ENGINEERING DESIGN**

PAMELA A. DREW, *Chair,* The Boeing Company
CAROL L.J. ADKINS, Sandia National Laboratories
GREGORY AUNER, Wayne State University
RON BLACKWELL, AFL-CIO
THOMAS W. EAGAR, Massachusetts Institute of Technology
ROBERT E. FONTANA, JR., Hitachi Global Storage Technologies
PAUL B. GERMERAAD, Intellectual Assets, Inc.
TOM HARTWICK, Adviser, Snohomish, Washington
ROBERT M. HATHAWAY, Oshkosh Truck Corporation
PRADEEP K. KHOSLA, Carnegie Mellon University
JAY LEE, University of Wisconsin, Milwaukee
DIANA L. LONG, Consultant, Charleston, West Virginia
MANISH MEHTA, National Center for Manufacturing Sciences
NABIL Z. NASR, Rochester Institute of Technology
ANGELO M. NINIVAGGI, JR., Plexus Corporation
JAMES B. O'DWYER, PPG Industries
HERSCHEL H. REESE, Dow Corning Corporation
H.M. REININGA, Rockwell Collins, Inc.
LAWRENCE J. RHOADES, Ex One Corporation
JAMES B. RICE, JR., Massachusetts Institute of Technology
DENISE F. SWINK, Adviser, Germantown, Maryland
ALFONSO VELOSA III, Gartner, Inc.
BEVLEE A. WATFORD, Virginia Polytechnic University
JACK WHITE, Altarum

**Staff**

TONI MARECHAUX, Director

# Preface

Today's defense systems incorporate an increasing number of electronic components, intended to enable these systems to be more accurate, more sophisticated, and more effective. Advances in printed circuits and associated interconnection—an integral technology—have enabled this trend, and these advances are expected to continue to enable future combat systems.

To examine a number of issues surrounding the manufacturing and supply of these components, the National Research Council convened a panel of experts—the Committee on Manufacturing Trends in Printed Circuit Technology—to examine trends in electronics interconnection technology and manufacturing and their effect on U.S. defense needs.

The charge to the committee was specifically to do the following:

- Examine worldwide and U.S. trends in technology investment and manufacturing competences for printed circuit boards.
- Assess the role of printed circuit boards in maintaining U.S. military capability, especially in meeting unique defense needs.
- Examine current laws, policies, and regulations that pertain to printed circuit board manufacturing and their impact on maintaining future military capability.
- Describe potential strategies for research, development, and manufacturing for printed circuit boards to meet both legacy and future U.S. defense needs.

A meeting was held December 13 and 14, 2004, attended by committee members, expert consultants, and Department of Defense (DoD) representatives. Technical topics were presented and discussed covering the general areas of system considerations, the suitability of current supply practices, the influence of new technologies, and technology insertion. DoD representatives provided a useful overview and rationale to set the stage for the discussions. Formal presentations were brief in order to allow for significant interactions between committee members and guests to home in on responses to the tasks listed above. After the meeting, the committee continued to gather information and to discuss and deliberate on findings, conclusions, and recommendations.

This report has been reviewed in draft form by individuals chosen for their diverse perspectives and technical expertise, in accordance with procedures approved by the National Research Council's Report Review Committee. The purpose of this independent review is to provide candid and critical comments that will assist the institution in making its published report as sound as possible and to ensure that the report meets institutional standards for objectivity, evidence, and responsiveness to the study charge. The review comments and draft manuscript remain confidential to protect the integrity of the deliberative process. We wish to thank the following individuals for their review of this report: Doug Freitag, Bayside Materials Technology; Steven P. Gootee, SAIC; Carol Handwerker, Purdue University; R. Wayne Johnson, Auburn University; Paul G. Kaminski, Technovation, Inc.; Robert Pfahl, iNEMI; Joe Schmidt, Raytheon; and Frank Talke, University of California, San Diego.

Although the reviewers listed above have provided many constructive comments and suggestions, they were not asked to endorse the conclusions or recommendations, nor did they see the final draft of the report before its release. The review of this report was overseen by Elsa Garmire, Dartmouth College. Appointed by the National Research Council, she was responsible for making certain that an independent examination of this report was carried out in accordance with institutional procedures and that all review comments were carefully considered. Responsibility for the final content of this report rests entirely with the authoring committee and the institution.

The committee also acknowledges the speakers from government and industry who took the time to share their ideas and experiences. H.M. Reininga, Board on Manufacturing and Engineering Design liaison to the committee, also greatly assisted the work of the committee through his participation in many of the committee's activities. Finally, the committee acknowledges the contributions to the completion of this report from the staff of the National Research Council, including Marta Vornbrock, Laura Toth, and Toni Marechaux, as well as those of Albert Alla, an intern at the National Research Council who assisted in background research for the report.

David J. Berteau, *Chair*
Committee on Manufacturing Trends in Printed Circuit Technology

# Contents

# Tables, Figures, and Box

## TABLES

## FIGURES

## BOX

# Summary

Today, many in the Department of Defense (DoD), the U.S. Congress, and the federal government lack a clear understanding of the importance of high-quality, trustworthy printed circuit boards (PrCBs) for properly functioning weapons and other defense systems and components. This report of the National Research Council's Committee on Manufacturing Trends in Printed Circuit Technology aims to illuminate the issues related to PrCBs for military use. In addition, this report offers recommendations that will help DoD to (1) preserve existing systems' capabilities, (2) improve the military's access to currently available PrCBs, and (3) ensure access to future PrCB technology. The recommendations reflect the need to achieve these goals at reasonable cost and with due respect for evolving environmental regulations.

To some, PrCBs may seem an older technology, declining in use for cutting-edge weapons systems and defense technology. In fact the opposite is true. PrCBs connect, in increasingly sophisticated ways, a variety of active components (such as microchips and transistors) and passive components (such as capacitors and fuses) into electronic assemblies that control systems. Given the military's increasing interest in and reliance on networked operations, these applications will expand for the foreseeable future, and the use of and requirements for PrCBs will continue to grow. While many of those requirements can be satisfied by commercial components, significant defense needs will be met only by the production of specialized, defense-specific PrCBs that are unavailable from commercial manufacturers.

The effectiveness of defense systems depends on the underlying PrCB technology. This report addresses several key related concerns raised by the committee. These include (1) access to PrCBs and PrCB technologies that can meet defense-related requirements, (2) the overall reliability of the PrCBs themselves, (3) the vulnerability of the PrCB supply chain to disruption, and (4) the secure operation of defense systems for which PrCBs are a component. Since PrCBs are essential to defense systems, these considerations have to be addressed so that defense-critical PrCBs can be protected from tampering and so that access to them can be assured. Without these assurances, systems may not work as planned in support of DoD's missions. When DoD uses suppliers of PrCBs that are trusted domestic sources, these considerations are easier to address than when the sources and distribution are global, as is increasingly the case for PrCBs. The solutions thus require an understanding of defense needs and DoD policies as well as the global market and its trends. This report develops those understandings.

## THE CURRENT SITUATION

Three major factors combine to affect the current situation for defense PrCBs. First, over the past two decades, DoD policy has led to a reduction in defense-specific manufacturing and a parallel increase in support for commercial-military integration by industry. Thus, DoD policy is to rely for the procurement of defense system components on the commercial sector wherever possible. For legacy systems already in the DoD inventory, DoD policy relies on a combination of private sector businesses and DoD-owned capability to sustain performance through maintenance, repair, and necessary upgrades.

Second, during the same period, the U.S. domestic PrCB industry has undergone two major alterations. It has changed from one in which U.S. production dominated (with 42 percent of global revenue in 1984) to one in which U.S. production is projected to be less than 10 percent of global revenue in 2006.[1] In addition, the mix of PrCB products has shifted because of increasing consumer use. Today, more than half of the PrCBs produced worldwide are for high-volume, low-cost, short-lived products such as cellular telephones, small appliances, and toys.

Third, many DoD requirements have become more sophisticated. Applications generally call for long life in PrCBs, with performance on demand under extreme conditions, with very high reliability. These requirements cannot be met by high-volume, short-lived consumer products. In fact, few if any defense-specific components with such characteristics can even be provided by manufacturers of PrCBs used in commercial durable goods such as automobiles, appliances, and heavy equipment, because of the high cost of interrupting high-efficiency production to manufacture a handful of defense-unique PrCBs.

As a result of these three factors, PrCBs for consumer products, commercial goods, and defense systems are increasingly manufactured by different companies that have little overlap in processes or products. Thus, DoD's policy to procure from commercial manufacturers is becoming difficult to implement for many PrCB applications.

This situation is complicated by an additional policy concern. When DoD program managers buy weapons systems, the focus is on the best price for purchase of the total system, not the reliability and trustworthiness of individual components such as PrCBs. Under this policy, there is currently little incentive for or ability to justify spending more to ensure that individual defense system components like PrCBs will perform reliably and be protected from tampering during their manufacture, assembly, and distribution. Absent funding that allows for such concerns, little effort can usually be allocated to assessing the sources of supply for PrCB components or subcomponents. However, well-developed mechanisms for improving supply-chain management are available, if program managers were directed by policy to pursue better reliability and performance of defense system components.

An additional challenge exists, even if current production considerations are resolved through policy and funding changes. Defense requirements change continuously, and DoD needs to ensure access to sufficient innovation to continue to meet new defense needs for improved PrCBs. DoD has traditionally stimulated innovation to meet emerging requirements by directly funding research and development (R&D) contracts or by reimbursing defense contractors for their own R&D costs. This approach worked well in the early days of electronics, but in the case of PrCBs today, even the global defense business base is not large enough to sustain that approach. What will be the source of that needed innovation?

Commercial-military integration policy relies on the commercial market to meet defense needs. However, commercial manufacturers' capacity for and spending on R&D has declined, and the remaining limited technology innovation is targeted at high-volume consumer goods. While this approach may support some DoD needs, such innovation will have little applicability in supporting and enhancing high-performance defense-related systems' capability. In addition, the long design and procurement cycles for DoD systems (often lasting more than a decade) lead to a fundamental disincentive both for developing and for adopting new technologies for defense applications. The result has been a steady decrease in innovation in DoD systems, even in programs with funding levels once considered reasonable and adequate for this purpose.

The continuing vitality of both the commercial domestic manufacturing sector and the global defense sector depends on three elements: (1) sources of research and technology developments, (2) innovation in the supply base for materials and chemicals, and (3) the availability of a skilled workforce. DoD must address all three elements to remain innovative and successful.

Both for current defense systems and for future technology, DoD needs the right blend of commercial innovation, defense incentives, and funding. What is currently not known is whether that blend can be identified and put in place to encourage a reliable supply of high-quality PrCBs for defense systems. Perhaps more importantly, there is at present no clear understanding of the fit between DoD-specific needs for PrCBs and the corresponding commercial industrial capabilities for meeting those

---

[1] E. Henderson. 2005. PCI Market Research Service Report. Los Altos, Calif.: Henderson Ventures.

needs and no clear definition of specific investments that might yield results that meet the needs of current and future defense systems.

## FINDINGS AND CONCLUSIONS

DoD is crucially dependent on the ability to support currently fielded systems made up of older components, known as legacy systems. Many of these systems contain PrCBs that are several generations behind today's off-the-shelf production. The committee found that existing small-firm contractors and DoD in-house capability are likely to be sufficient to sustain legacy systems, although that capability will need regular funding in order to maintain efficient manufacturing technology for repairing or replacing older PrCBs.

For current and especially for future applications of PrCBs, the committee found that there is currently no adequate set of information or paradigm for DoD to use in determining what is needed to ensure adequate access to reliable and trustworthy PrCBs for use in secure defense systems. So that such a body of information can be developed and put to use, the committee recommends an approach that would also be applicable to specific areas of concern, such as the transition of PrCB technology and products to meet lead-free standards. More specifically, the committee calls on a variety of experts to review the following three areas:

- The need for an existing PrCB component or new PrCB technology should be assessed by military planning groups, and the results used to ensure access to the technologies required to field effective defense systems.
- The vulnerability of a defense system attributable to the PrCB component will require a separate assessment of operational characteristics and performance as well as potential exposures to security risks in the supply chain. The resulting information should be used to ensure the reliability and trustworthiness of PrCBs for secure, effective defense systems.
- The threat potentially posed to overall defense capabilities by lack of access to high-quality, trusted PrCB component technology will require a more specialized assessment for understanding how best to use DoD resources to maintain and enhance the nation's security.

DoD is capable of addressing all three of these areas, but it does not now do so in a systematic manner. The results of such reviews could help enable the federal government and the defense industrial base to work together to preserve and build critical systems whose underlying trusted PrCB component technologies ensure desired performance capabilities, with the ultimate goal of ensuring continuity of supply and adequate security. Assessments such as those called for by the committee will also allow DoD to deal with such emerging trends as the global migration to lead-free PrCB technology.

## RECOMMENDATIONS

**Recommendation 1:** The Department of Defense should address the ongoing need for printed circuit boards (PrCBs) in legacy defense systems by continuing to use the existing manufacturing capability that is resident at the Naval Surface Warfare Center, Crane Division (Indiana) and at Warner Robins Air Logistics Center (Georgia), as well as contractors currently providing legacy PrCB support.

**Recommendation 2:** The Department of Defense should develop a method to assess the materials, processes, and components for manufacture of the printed circuit boards (PrCBs) that are essential for properly functioning, secure defense systems. Such an assessment would identify what is needed to neutralize potential defense system vulnerabilities, mitigate threats to the supply chain for high-quality, trustworthy PrCBs, and thus help maintain overall military superiority. The status of potentially vulnerable materials, components, and processes identified as critical to ensuring an adequate supply of appropriate PrCBs for defense systems should then be monitored.

**Recommendation 3:** The Department of Defense (DoD) should ensure its access to current printed circuit board (PrCB) technology by establishing a competing network of shops that can be trusted to manufacture PrCBs for secure defense systems. In addition to being competitive among themselves, these suppliers should also be globally competitive to ensure the best technology for the U.S. warfighter and should be encouraged and supported to have state-of-the-art capabilities, including the ability to manufacture PrCBs that can be used in leaded and lead-free assemblies. To maintain this network of suppliers, DoD should, if necessary for the most critical and vulnerable applications, purchase more PrCBs than are required to meet daily consumption levels in order to sustain a critical mass in the trusted manufacturing base.

**Recommendation 4:** The Department of Defense (DoD) should ensure access to new printed circuit board (PrCB) technology by expanding its role in fostering new PrCB design and manufacturing technology. DoD should sponsor aggressive, breakthrough-oriented research aimed at developing more flexible manufacturing processes for cost-effective, low-volume production of custom PrCBs. In conjunction with this effort, DoD should develop explicit mechanisms to integrate emerging commercial PrCB technologies into new defense systems, even if that means subsidizing the integration. These mechanisms should include more innovative design capabilities and improved accelerated testing methods to ensure PrCBs' lifetime quality, durability, and compliance with evolving environmental regulations for the conditions and configurations unique to DoD systems.

The committee believes that taking these recommended steps will help DoD to preserve its legacy defense systems, meet current system requirements, and provide for future PrCB technology advances efficiently and securely. DoD needs no less than these outcomes to maintain U.S. military capability for the foreseeable future.

# 1

# Background and Overview

The function of a printed circuit board (PrCB), simply, is to connect a variety of active components (such as microchips and transistors) and passive components (such as capacitors and fuses) into an electronic assembly that controls a system. A typical printed circuit board consists of conductive "printed wires" attached to a rigid, insulating sheet of glass-fiber-reinforced polymer, or "board." The insulating board is often called the substrate.

An important characteristic of PrCBs is that they are usually product-unique. The form factor— meaning the size, configuration, or physical arrangement—of a PrCB can range from a system literally painted on to another component, to a structural element that supports the entire system.

The first PrCBs made on a large scale were manufactured in 1943 when the U.S. military began to use the technology to make rugged radios for use in World War II.[1] Originally, individual devices were attached to an interconnecting medium called a board, which was usually produced by the same company that made the system. In the 1970s and 1980s, PrCBs were commoditized for a specialty market. Today, markets for this interconnection technology range across the whole of the global economy, and include the following areas:

- Government, military, and aerospace uses;
- Medical devices;
- Automotive electronics;
- Computers and business electronics;
- Consumer electronics;
- Industrial electronics and instrumentation; and
- Communication.

Today, interconnecting electronics in increasingly complex systems is leading to complex designs, components, and systems. The advent of integrated electronics, such as a system-on-a-chip and multichip modules, has increased speed and reduced latency in electronics. The interconnections for these components have become equally diverse.

As is shown in Figure 1-1, many PrCBs play a dual role in products— both serving as a structural element and performing an electrical function. Because of these complexities, their manufacturing process is also complex. Contributors to the final PrCB product include designers, board manufacturers, assembly companies, suppliers, and original equipment manufacturers (OEMs). Appendix F illustrates and describes the fabrication steps for a standard PrCB, and the following sections give more details on the ingredients for this fabrication.

---

[1] Wikipedia. Printed circuit board. Available at http://en.wikipedia.org/wiki/Printed_circuit_board. Accessed October 2005.

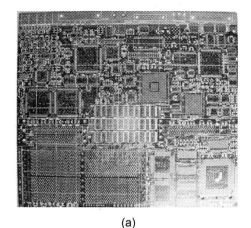

(a)

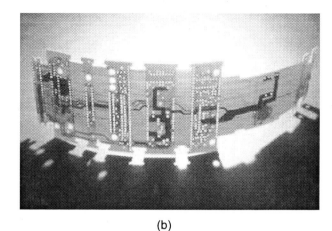

(b)

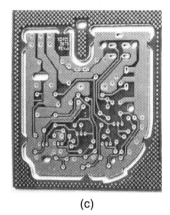

(c)

FIGURE 1-1 An array of printed circuit boards in various sizes, form factors, and materials. (a) A rigid 18-layer board for computer applications; (b) a flex board for cellular telephone applications; and (c) a rigid 2-layer board for automotive applications. SOURCE: CALCE Electronic Products and Systems Center, University of Maryland, and IPC, Association Connecting Electronics Industries.

## BOARD MATERIALS

One important degree of complexity in the manufacture of PrCBs is entailed in the base material, or combination of materials, of the board. An astonishingly broad range of materials and form factors are used, and are often combined in many different ways. For example, the interconnect circuit may be painted onto other components, or the board may have polymer, glass, or ceramic substrates.

For example, many boards are not very boardlike in that they are neither rigid nor thick—simple PrCB substrates, for example, can be a paper-based laminate impregnated with phenolic resin. This type of board carries designations such as XXXP, XXXPC, or FR-2. The material is inexpensive; it is easy to machine by drilling, shearing, or cold punching; and it also causes less tool wear than that resulting from glass-fiber-reinforced substrates. The letters FR in the designation indicate flame resistance.

Higher-end circuit board substrates for industrial or selected commercial applications are typically made of the material designated FR-4. This is a woven fiberglass mat impregnated with a flame-resistant epoxy resin. It can be drilled, punched, and sheared, although the abrasive quality of the glass reinforcement requires tungsten carbide tooling for high-volume production. The fiberglass gives this material much higher flexural strength and resistance to cracking than paper-phenolic types of boards have, but at a higher cost.

PrCBs for high-power radio-frequency (RF) applications require plastics with low dielectric constant (permittivity) and dissipation factor, including polyimide, polystyrene, polytetrafluoroethylene, and cross-linked polystyrene. These typically trade off mechanical properties, such as strength and lightness, for superior electrical performance. Another specialty application of PrCBs is their design for use in vacuum or in zero gravity, as in spacecraft, in conditions that preclude reliance on convection cooling. These PrCBs often have thick copper or aluminum cores to dissipate heat from their electrical components.

Not all circuit boards use rigid core materials. Some are designed to be completely flexible or partially flexible, using polyimides or other films. Boards in this class, sometimes called flex circuits or rigid-flex circuits, can be more difficult to produce but have many applications. Flexibility can save space in applications such as cameras and hearing aids. Also, a flexible part of a circuit board can serve as a connection to another board or device. Some boards may also combine rigidity and flexibility—for example, the cable connected to the carriage in an inkjet printer.

Boards can be one-sided or two-sided, they can have metallic or nonmetallic vias (holes connecting different layers of circuitry), they can be multilayered with different structures on different levels, and so on. Printed boards may be classified according to different base materials and different structures, sometimes both. Examples include one-sided phenolic aldehyde paper-base printed boards and multilayer polyimide printed boards.

## BOARD DESIGN

The main function of printed boards is to support and interconnect the electronic components mounted on them; they may also serve to dissipate heat and protect components. The base materials, wires, and wire layers vary widely; design decisions are made according to the specific requirements of the application. Constraints include the size, weight, and shape of the substrate, because most assemblies are designed to support the components and to be a structural component. Other constraints include considerations involving power needs, heat generation and dissipation, severity of service use, efficiency, reliability, and cost.

In some designs, the electronic components mounted on a board can be viewed as simple building blocks that are controlled by programmable software, with the board containing the logic of the system. In other designs, the board can be simple, the components carry the brains, and little software is needed. In supercomputers, for example, both the chips and boards are relatively simple; in such a case, many of each are tied together in their computing purpose through sophisticated software.

These trade-offs in design provide a broad array of challenges for subsystem and system integrators. Many times, design parameters for a subassembly are set by the design of the larger assembly that will use it. At other times, design choices are driven by previous experience of the designer company, or by the design software, or manufacturing equipment available, or component availability.

These external drivers for system design can become more important than considerations of simple cost or ease of configuration. For example, very different constraints apply to high-volume, low-mix components than to highly specialized, low-volume designs. Design decisions can also be tied directly to the overall security of the manufacturing process and the supply logistics of the final system. OEMs are in the early stages of understanding and managing these trade-offs.

An additional overriding issue in design can be concern for where to locate the "brains" of the system. The intelligent components carry the logic and can also carry valuable intellectual property. Therefore, the potential for copying, counterfeiting, or subverting a component, and possibly an entire system, must be considered. A system with complex hardware, software, and interconnections could allow the possibility of a coordinated subversion that could be impossible to detect.[2]

## MANUFACTURING TRENDS

Manufacturing in the United States has traditionally been a strong sector of the economy, contributing 20 to 30 percent of the gross domestic product (GDP).[3] Manufacturing in the United States is estimated to generate two-thirds of the nation's research and development and three-fourths of its exports and to support more than 20 million jobs. According to the National Association of Manufacturers, "Today,

---

[2] Defense Science Board. 2005. High Performance Microchip Supply. Washington, D.C.: Department of Defense Office of the Under Secretary of Defense for Acquisition, Technology, and Logistics.

[3] Information available at http://www.manufacturing.gov. Accessed October 2005.

manufacturing output, efficiency, and productivity are at record levels, capital investment is rising, and product quality has never been higher."[4]

Manufactured products have been an integral and fundamental component of the U.S. economy; they include goods such as analytical equipment to improve health care, computers and peripherals to power the information age, advanced weapons to promote defense, and a wide variety of vehicles to move the transportation industry forward. Manufacturing in many ways provides the substance for our quality of life and ability to advance as a nation.

Recent attention to the value of manufacturing to the nation by lawmakers and government agencies has reinforced this view. According to the Assistant Secretary of Commerce for Manufacturing and Services, "Manufacturers are full partners in the effort to build the future of the country in the marketplace for new products and ideas. Simply put, a healthy manufacturing sector is key to better jobs, fostering innovation, rising productivity, and higher standards of living in the United States."[5]

Some basic manufacturing procedures are shared by all PrCBs, although different technologies and equipment are used in the process. The particular technologies and equipment used are based on a number of factors, including the following:

- The thickness and quality of the base material;
- The width of the wire on printed boards;
- The width between wires and the resolution of their spacing;
- The routing density, which drives layer count and hole size;
- The structure of the printed boards;
- The manufacturing scale;
- Projected assembly techniques;
- Specific requirements made by customers; and
- Any special techniques used in manufacturing.

Because the technology—as well as the equipment used to implement it in printed board manufacturing—develops rapidly, production enterprises find it necessary to add to or update their techniques and equipment regularly, and often annually. The cost of equipment and the need to update create a gap between large-scale enterprises and smaller businesses that build to stringent product qualifications; the difference is revealed by their relative investment in continuous technology innovation. The fact that small-scale enterprises cannot invest as readily affects their ability to innovate and eventually also limits their need for technology innovation because they become bound to a limited market. Some top manufacturers, with large-scale, high-value, or complex processes, may invest between $20 million and $50 million per year.

The reasons that the PrCB industry is so technology-intensive and capital-intensive are numerous. They may include the following:

- *Various sophisticated processes are needed.* The manufacture of PrCBs includes work in the areas of optics, automatic control, electronic controls, intelligent processing, and electrochemistry.
- *Many techniques are involved.* These may include computer-aided design and computer-aided manufacturing, optical image transfer, high-speed and laser drilling, dielectric metallization, copper electroplating, tin electroplating, acid and alkaline etching, nickel and gold electroplating, laser direct imaging, hot-air leveling for final finish metals such as tin, liquid photoimageable resists, vacuum or autoclave lamination for multilayer products, automated x-ray systems for registration of layers, flying probe and compliant pin electrical testing, and automated optical inspection.

---

[4] National Association of Manufacturers. 2005. Pro-Growth and Pro-Manufacturing Agenda. Washington, D.C.: National Association of Manufacturers. Available at http://www.nam.org/s_nam/doc1.asp?CID=4&DID=232739. Accessed October 2005.

[5] Testimony of Albert A. Frink, Assistant Secretary of Commerce for Manufacturing and Services, before the Subcommittee on Technology, Innovation, and Competitiveness of the Committee on Commerce, Science, and Transportation, U.S. Senate, June 8, 2005. Available at http://commerce.senate.gov/hearings/testimony.cfm?id=1526&wit_id=3678. Accessed October 2005.

- *Numerous procedures are involved.* As many as 30 or 40 procedures are needed in the manufacturing of multilayer boards; often one procedure consists of more than 10 individual steps.
- *Highly specialized equipment is required.* Most of the processing equipment and the manufacturing tools sets are automated, computer-controlled, or programmable-logic-controlled (PLC) systems designed to provide the high level of accuracy needed for the fabrication of a PrCB. The specialized equipment set includes laser photo plotters; PLC chemical processing lines; numerically controlled devices; hot oil, electric, or autoclave lamination presses; automated optical inspection systems; automated exposure devices; roller or screen coating systems for dielectric applications; and multilayer registration tools.
- *Many different types of materials are needed.* More than 100 different materials are used in the manufacturing process for most PrCBs. Some of these materials become a part of the PrCB, including the copper-clad laminate materials consisting of copper films, epoxies, or other dielectrics, with the addition of reinforcements such as fiberglass in some cases; the electroplated metals; the solder mask dielectric materials; and the metallic or organic final finish used to improve the assembly and soldering processes. Other materials have a specific use during processing and are discarded after use. Such process consumables include photosensitive dry films or liquid resists, special-purpose adhesive tapes, stop-off agents, fluxes, acids, bases, cleaners, and etches. These process consumables and the wastes produced must also be disposed of properly.
- *Careful control of the manufacturing environment must be maintained.* In addition to the rigorous requirements for the equipment sets used in the manufacturing process, there is a need for rigorous control of the manufacturing environment in terms of cleanliness, temperature, and humidity. Photolithographic and lamination buildup process areas are often environmentally closed work areas. Class 10,000 (and even class 1,000) clean rooms with rigid temperature and humidity control are commonly used in the photolithography areas in particular.[6]

Beyond the issues described above, it is important to note that access to printed circuit technology is essential to manufacturing know-how for all electronics in the U.S. economy. Semiconductor technology performance continues to double every 18 months,[7] and most semiconductor chips require packaging that includes some form of interconnecter such as a printed circuit board.

The increasing globalization of the electronics industry has driven the capability to manufacture interconnection technology overseas.[8] The intense competition in the face of this increasing globalization currently challenges U.S. manufacturers and leaves many U.S. firms unable to raise prices to keep pace with rising production costs. Without a technology innovation base, they are also unable to increase their productivity.

This is a key challenge for the domestic PrCB industry. Because PrCBs are not end products but intermediate products, the location of partner manufacturers is important. Many of the markets, or downstream customers, for electronic systems are moving or have moved overseas. In addition to facing a diminishing domestic market, U.S. PrCB manufacturers that look for global markets may find it difficult to compete in foreign markets that are insular with respect to U.S. producers. To be successful, companies must follow their markets offshore, which eventually could leave a base too small to support U.S. defense needs.

Despite the promise of a truly global free-trade scenario, the continued dissipation of downstream electronic systems components manufactured in the United States inevitably means that the Department

---

[6] A clean room is a work area in which the air quality, temperature, and humidity are highly regulated in order to protect sensitive equipment from contamination. Clean rooms are rated as "Class 10,000" if there are no more than 10,000 particles larger than 0.5 microns in any given cubic foot of air. "Class 1,000" clean rooms are ones in which there exist no more than 1,000 particles.

[7] G.E. Moore. 1965. Cramming more components onto integrated circuits. Electronics 38:114-117. While true at present, the trend may be slowing as the constraints in solid-state physics become increasingly difficult to overcome without fundamental advances in new technologies.

[8] T. Friedman. 2005. The World Is Flat. New York: Farrar, Straus, and Giroux.

of Defense will have less access to and availability of leading-edge electronic subsystem technology including PrCBs, microchips, and displays.[9]

## EVOLVING ROLE OF PrCBs

For many years, the manufacturing of PrCBs was in the category of commodity manufacturing and was carried out by vertically integrated companies that manufactured electronic equipment. However, as modern techniques have been developed, products have undergone dramatic diversification and specialization. And as the production scale and the required investment for PrCBs have grown, dedicated enterprises have emerged. The industries for manufacturing many of the materials and components contributing to today's PrCB have also become specialized.

Some estimates for the calendar year 2003 help place the industry in an overall context:[10]

- World GDP                                                    $49 trillion
- U.S. GDP                                                      $10.4 trillion
- Worldwide spending on information technology                  $2.3 trillion
- Worldwide electronic equipment sales                          $1.1 trillion
- U.S. Defense spending                                         $405 billion
- U.S. Defense electronics spending                            $75 billion
- Worldwide PrCB sales (rigid)                                  $29 billion
- U.S. PrCB sales                                               $4.4 billion
- U.S. PrCB defense spending                                   $500 million

A major factor differentiating PrCBs from other electronic components is that PrCBs are wholly customized components. This means that products must be made according to specific designs, characteristics, quantity, and delivery schedules. The generally low margins for commodity components are difficult for PrCB manufacturers to meet for a number of reasons, including the use of a variety of materials with a limited shelf life, the variety of possible trade-offs between design and manufacturing processes, and the many different potential processes and combinations of processes. These factors make the specialty manufacturing of PrCBs a higher-cost proposition, whereas economies of scale can enable the delivery of PrCBs at low cost for some consumer products. These constraints also mean that survival in the industry necessitates very tight management of processes and process controls.

---

[9]  S. Cohen and J. Zysman. 1987. Manufacturing Matters: The Myth of the Post-Industrial Economy. New York: Basic Books.

[10] These are estimates only and are sourced from a number of publications that may have used different underlying assumptions and definitions. Sources included the 2003 CIA World Fact Book; the Government Electronics and Information Association 15th Annual Forecast; the Information Technology Association of America; the Congressional Budget Office Summary Update for Fiscal Year 2003; and IPC, the Association Connecting Electronics Industries. The committee realizes that many additional sources for such data are available via an Internet search and that the error in these numbers may be 50 percent or more. The data are intended only to provide a frame of reference.

# 2

# The Printed Circuit Technology Industry

Very few generalizations can be made about the global printed circuit board (PrCB) industry, other than to say that it is diverse. While many concerns are shared by the many manufacturers, suppliers, technologists, and traders that make up the printed circuit industry, many of these same parties also have conflicting beliefs on whether current trends are good or bad.

A parallel observation is that very few generalizations can be made about the markets for printed circuit technology other than to say that applications are ubiquitous and growing. A brief look around any environment will reveal dozens to hundreds of printed circuit boards in items ranging from garden tools to cellular telephones to high-end supercomputers. As varied as these applications are, a number of trends in the technology are expected to influence their use.

## INDUSTRY OVERVIEW

The global industry for the design and production of printed circuit technology is constantly evolving. The following is intended as a snapshot view of the industry in the United States. Table 2-1 shows the breakdown as of 2003 in the various types of boards produced and offers a broad view of U.S. and global production.

### Size of Market, Capacity, and Companies

In 2003, the dollar value of the U.S. PrCB market was approximately $4.4 billion, down more than $6 billion from 2000 when the dollar value of the market was approximately $10.7 billion. In 2000, the government and military segment of the market was 2 percent, or more than $200 million. In 2003, this segment had risen to 12 percent of the total and accounted for more than $500 million in sales.[1]

While it is difficult to determine capacity in light of the continuous cycle of PrCB manufacturing plant closures that is currently going on, estimates of capacity based on returning to round-the-clock operation of all U.S. facilities are suspect. No U.S. PrCB manufacturing location is reported to be running at 100 percent capacity—a trend that is expected to be long term owing to the capital expense, training, and retooling time needed for facilities to return to round-the-clock operations after 4 to 5 years of two-shift operation.

In 2000, 13 independent rigid-PrCB manufacturers in the United States each had sales of over $100 million annually; an additional 30 U.S. independent manufacturers each had sales of between $50 million and $100 million. These 43 companies were the backbone of the industry-wide research and development (R&D) effort in the United States and as such were willing to take risks and invest in new

---

[1] D. Bergman, IPC. 2004. Presentation to this committee. December 13.

TABLE 2-1  Dollar Value of Printed Circuit Board Production by Global Region in 2003 (millions of dollars)

| Region | Paper | Composite | Glass Epoxy | Multilayer Epoxy | Multilayer Nonepoxy | High Density MicroVia | Integrated Circuit Substrates | *Rigid PrCB Subtotal* |
|---|---|---|---|---|---|---|---|---|
| Asia/Pacific | 1,296 | 1,041 | 2,414 | 8,811 | 400 | 3,219 | 3,576 | *20,756* |
| Europe | 174 | 178 | 1,081 | 1,356 | 85 | 439 | — | *3,312* |
| Middle East and Africa | 2 | — | 62 | 54 | 10 | 5 | 5 | *138* |
| North America | 30 | 40 | 850 | 3,191 | 524 | 155 | 57 | *4,847* |
| Other Americas | 23 | 3 | 56 | 24 | — | 2 | — | *108* |
| World Total | 1,524 | 1,262 | 4,463 | 13,436 | 1,019 | 3,820 | 3,638 | *29,161* |

| | Flex Circuits | Rigid-Flex Circuits | *Flex Circuits Subtotal* |
|---|---|---|---|
| Asia/Pacific | 3,888 | 555 | *4,443* |
| Europe | 121 | 123 | *244* |
| Middle East and Africa | 5 | 10 | *15* |
| North America | 500 | 121 | *621* |
| Other Americas | 1 | — | *1* |
| World Total | 4,515 | 809 | *5,324* |

| | Grand Total PrCBs |
|---|---|
| Asia/Pacific | 25,199 |
| Europe | 3,555 |
| Middle East and Africa | 153 |
| North America | 5,468 |
| Other Americas | 109 |
| World Total | 34,484 |

SOURCE:  IPC, the Association Connecting Electronics Industries.

processes or equipment to improve the quality and technology of their products.[2] In total, 678 independent rigid-PrCB manufacturing companies operated in the United States in 2000. Table 2-2 categorizes the 678 companies according to their annual sales.

In 2003, only 8 independent rigid-PrCB manufacturers were operating in the United States with sales of over $100 million each; 5 companies had sales of between $50 million and $100 million each. This 61 percent decline in large, well-funded, and independent PrCB manufacturers (those with annual sales of over $50 million) is a contributing factor in the decline of technology innovation and investment in the United States. In total, it is estimated that fewer than 500 independent rigid-PrCB manufacturers remained in the United States in 2003, down 27 percent overall from the total of 678 in 2000. While some of this decline may be due to increased productivity that can lead to internal consolidation or consolidation through acquisition, the overall numbers are decreased across the board.

These data describe the exodus of PrCB manufacturing offshore during the period from 2000 to 2003.  For the PrCB industry, one of the greatest concerns was the loss of the larger independent manufacturers.  This segment was the most critical to the continuation of U.S. technology innovation and investment. It is unlikely that the companies that remain—most with sales under $20 million annually—will be able to make the investment required today and into the future to maintain competency in the state-of-the-art manufacturing practiced by the global leaders in Japan, Taiwan, and now rapidly emerging in China. Many industry consultants also believe that the remaining companies in the United States,

---

[2]  Most of these companies had been participating members of the Association Connecting Electronics Industries (known as IPC), an industry trade association, and the now-defunct Interconnection Technology Research Institute (ITRI).

TABLE 2-2  Number of Independent U.S. Companies Manufacturing Rigid PrCBs, 1995, 2000, and 2003

| Annual Sales per Company | 1995 | 2000 | 2003 |
|---|---|---|---|
| Over $100 million | 7 | 13 | 8 |
| $50 million to $100 million | 15 | 30 | 5 |
| $20 million to $50 million | 36 | 39 | 31 |
| $10 million to $20 million | 60 | 88 | 62 |
| $5 million to $10 million | 90 | 149 | 122 |
| Under $5 million | 460+ | 359+ | 265+ |
| Total | 668+ | 678+ | 493+ |

NOTE:  Plus sign (+) indicates a low-end estimate.  SOURCE: IPC, the Association Connecting Electronics Industries.

currently about 400, will have a difficult time competing in the global marketplace and may face competition even in niche products over the next 2 to 5 years.

Of the roughly 400 active independent rigid-PrCB manufacturers that existed at the start of 2005, only 18 are certified under MIL-PRF-31032. In the flex and rigid-flex segments, only 6 companies are qualified under MIL-PRF-31032.  Half of these companies are under $20 million in annual sales.[3]

A growing sector in electronics manufacturing is in the contract manufacturing of electronics manufacturing services (EMS).  EMS has become a major component of the electronics industry in the past 5 years.  This migration is less true for the PrCB industry.  Only one electronics manufacturing systems company, Sanmina-SCI, is vertically integrated and manufactures PrCBs.  The rest of the EMS industry buys raw PrCBs from suppliers, such as Tyco or Sanmina-SCI, and then populates the boards with components, tests them, builds them into full systems, performs system-level tests, and provides logistics and repair support.  So, while EMS companies are not a major manufacturer of PrCBs, they represent a significant share of PrCB company customers.

**The Global Nature of the Industry**

The PrCB industry, like the larger electronics industry, has always had a global component. Only in the past 4 years, however, has the U.S. manufacturing base faced a serious decline. The decline is continuing as the remaining larger companies close facilities and increase their investment in China and in other lower-cost manufacturing locations.

In 2000, the United States was second only to Japan in production of PrCBs, and these two countries had swapped the lead position at various times in previous years, Table 2-3.  Looking at the top 10 producers of PrCBs from each region gives a reasonable perspective on the entire marketplace.  In 2000, Japan's top 10 had 25 percent of global capacity, and the United States' top 10 had 21 percent.  In the previous year the United States' top 10 led the pack, followed by Japan's top 10 and Taiwan's top 10, with China's still holding the fourth position. In 2003, Japan's top 10 continue to dominate, holding 29 percent of the global market share, with China's top 10 leaping to second position at 17 percent of the global market share; the United States' top 10 follow with 15 percent, and Taiwan's top 10 lag at 13 percent market share.

Along with the regional shifts in output, a fundamental regional consolidation has occurred as well. Fewer countries have a producer in the global top 10, and there are no longer any outliers to be captured

---

[3]  MIL-PRF-31032 is the "Performance Specification Printed Circuit Board/Printed Wiring Board, General Specification for FSC 5998."  This specification establishes the general performance requirements for printed circuit boards or printed wiring boards and the verification requirements for ensuring that these items meet the applicable performance requirements. The intent of this specification is to allow the printed board manufacturer the flexibility to implement best commercial practices to the maximum extent possible while still providing product that meets military performance needs.  Available at http://www.dscc.dla.mil/Programs/MilSpec/listdocs.asp?BasicDoc=MIL-PRF-31032.  Accessed September 2005.  Note that effective December 31, 1998, MIL-PRF-31032 replaced MIL-PRF-55110 for new design applications. MIL-PRF-55110 is still in use for many legacy applications.

TABLE 2-3 Annual Sales for Top Ten Companies in Printed Circuit Industry, 2000 and 2003

| Top 10 Producers, 2000 | | Million U.S. $ | Top 10 Producers, 2003 | | Million U.S. $ |
|---|---|---|---|---|---|
| 1  Sanmina-SCI | United States | 1,500 | 1  Nippon Mektron | Japan | 1,117 |
| 2  Visasystems | United States | 1,250 | 2  CMK | Japan | 1,049 |
| 3  CMK | Japan | 1,112 | 3  Ibiden | Japan | 1,027 |
| 4  Ibiden | Japan | 1,083 | 4  Hitachi Group | Japan | 685 |
| 5  Hitachi Group | Japan | 973 | 5  Shinko Denki | Japan | 636 |
| 6  Nippon Mektron | Japan | 905 | 6  Unimicron | Taiwan | 609 |
| 7  Compeq | Taiwan | 802 | 7  Samsung E-M | Korea | 545 |
| 8  Tyco | United States | 780 | 8  Compeq | Taiwan | 462 |
| 9  Fujitsu | Japan | 624 | 9  Nanya PCB | Taiwan | 453 |
| 10  Multek | United States | 600 | 10  Daeduck Group | Korea | 422 |
| Total | | 9,629 | | | 7,005 |

SOURCE: IPC, the Association Connecting Electronics Industries.

as "rest of the world." The result is that all of the top 10 manufacturers are from a key location in Asia, the United States, or Germany.

According to the PCI Market Research Service report of March 2005, "Irresistibly low costs and access to huge markets have created a Chinese magnet for PrCB producers, worldwide. The competitive fallout is evident in the continuing closures of facilities in North America and Europe. And recent announcements indicate that the process will continue in 2005. As a result, China will overtake Japan as the number-one board producer in 2006. That year, China is forecast to produce $10.6 Billion worth of PrCBs, accounting for 25 percent of the world total."[4]

One difficulty in reconciling this information is that the top three U.S. producers of PrCBs all have significant manufacturing bases outside the United States even though their annual sales are attributed to the United States. Of the top 25 PrCB manufacturers worldwide in 2003, only 4 were U.S. companies—Viasystems Group (11 plants), Sanmina-SCI (13 plants), Multek (14 plants), and Tyco PRCB (16 plants). Of these 4 companies, only one (Tyco PRCB) did not have a significant component of the production in offshore manufacturing. Of the top 10 PrCB manufacturers headquartered in the United States, half place the majority of their production in Asia.[5]

In 2000, nearly 80,000 people were employed in the North American PrCB industry. These jobs ranged from employment of hourly production workers to that of salaried engineers and included management, information technology, and professional workers. At the beginning of 2004, the total dropped to just over 41,000, nearly a 50 percent reduction in the labor force. These cuts were made across all job descriptions as plants closed, either owing to bankruptcy or because factories were relocated to Asia. No major technologies change was introduced during this period to increase productivity, so the decrease can be almost wholly attributed to production moved from U.S. to overseas locations.[6]

The number of people employed in North America in the PrCB industry continues to decline as new closures are announced weekly. In April 2005, Noble Industries, a manufacturer in Hibbing, Minnesota, closed its last plant—a 34,000-square-foot facility—having closed a Texas facility in 2001 and an Iowa plant in 2003.[7] Announcing another losing quarter, Sanmina-SCI stated that in 2004, "in response to market needs, we transferred manufacturing capacity from North America and Western Europe to lower-

---

[4]  E. Henderson. 2005. PCI Market Research Service Report. Los Altos, Calif.: Henderson Ventures.
[5]  IPC, the Association Connecting Electronics Industries.
[6]  IPC, the Association Connecting Electronics Industries.
[7]  K. Grinsteinner. 2005. 60 Hibbing workers lose jobs; Noble Industries closing down. Mesabi Daily News, April 27.

cost regions in Asia, Eastern Europe and Latin America."[8] Sanmina-SCI purchased Pentex Schweitzer in 2004, and in doing so, bought PrCB manufacturing capacity with one plant in Singapore and two facilities in China.[9] In early May 2005, the DDi Corporation announced the closure of its mass lamination facility in Phoenix, Arizona.[10]

Few bright spots are apparent in this picture. Older workers from the PrCB labor force in the United States may be forced to take either early retirement benefits or significant reductions in pay while they try to shift career paths. Few U.S.-based jobs are advertised in PrCB research, development, or product engineering, and the majority of the current approximately 30,000 workers will reach retirement in the 2015 to 2020 time frame. Current industry trends will make it difficult for corporate leadership to attract a future talent pool to continue to serve the industry's requirements.

The PrCB industry is the apparent victim of a fundamental transformation of the global electronics industry, which is characterized by very rapid product cycles and extremely demanding cost pressures and is led by very high volume throughput applications. For several reasons, much of the production capability for these downstream electronic products has increasingly moved offshore and increasingly to Asia.[11]

This inexorable trend has had major ripple effects in upstream supply, with PrCBs being only one of the affected industries. For the U.S. PrCB industry, the combination of low operating margins and low sales volume produces little or no investment in research, technology, or innovation.

For the Department of Defense (DoD) and for national security, the impact that the changes in the PrCB industry, as described above, have had on U.S. policy interests falls into two key categories. First, with greater emphasis on highly integrated electronic systems in which many functionalities are combined in a single device, advanced packaging and interconnection technologies become increasingly important, whether for consumer applications, industrial applications, or defense applications. Second, for highly specialized needs for defense, the access to both new technology and trusted production sources is endangered. Under current conditions, it is unlikely that technical capabilities, including a skilled workforce, can be sustained.

## High-Performance-Board Production

High-performance boards are those made primarily for military and medical applications. For the Department of Defense, qualification of suppliers for current production is done using MIL-PRF-31032.[12] High-performance boards require different and more-sophisticated equipment to be used in the manufacturing process. These equipment sets include but are not limited to laser drills, autoclave lamination presses, laser direct-imaging machines, laser trimmers, and specialized plating lines. When purchased new, these equipment sets have a high capital cost per unit; for example, the total cost for the equipment set just outlined is $4.4 million; it would provide only a very limited production output. In aggregate, a company with $10 million worth of process equipment would need to make an investment of more than $3 million per year to maintain state-of-the-art competency of the equipment, or about a 30 percent capital investment per year. In most cases in which investments have been delayed because of the downturn during the past 3 years and longer, companies would need to make a capital investment of more than 50 percent of their total revenues to return their manufacturing capabilities to current production.

As a result of the equipment intensity of this type of manufacturing, producing high-performance boards such as are required to meet government and military requirements, even in low volumes, would mandate a high capital investment by the existing U.S. manufacturing base. For independent PrCB manufacturers with sales of under $20 million annually, the possibility is very unlikely. It becomes reasonable to conclude that the 400 or so U.S. companies cannot hope to remain competitive in this high-

---

[8] Executive Letter to Shareholders in the Sanmina-SCI 2004 Annual Report.

[9] Press Release. 2004. Sanmina-SCI to Acquire Pentex-Schweizer Circuits Limited. Available at http://www.sanmina.com/pressroom/2004/062804.pdf. Accessed September 2005.

[10] Form 10-Q for DDi Corporation, filed August 10, 2005. Available at http://biz.yahoo.com/e/050809/ddic10-q.html. Accessed September 2005.

[11] K. Pildal. 2004. Asia's PCB Manufacturing: Dramatic Growth . . . and Decline. Circuitree, February. Available at http://www.circuitree.com. Accessed October 2005.

[12] For systems specified prior to December 1998, the preferred military specification is MIL-PRF-55110.

technology area without a significant infusion of capital. It is probable that without outside support, these small PrCB suppliers will not continue to be able to meet the requirements of U.S.-manufactured PrCBs for government and military applications.

Approximately 5 percent of PrCB industry manufacturers are military-qualified under the military specification currently in force. It is important to realize, however, that military boards can be manufactured under previous defense specifications or by nonmilitary certified processes. For example, some shops may make boards to IPC-6012 class 3, using the Single Process Initiatives acquisition excellence program. Boards specified prior to 1998 are under a number of older specifications. Table 2-4 shows the companies currently qualified to supply military boards under MIL-PRF-31032. Note that there are more qualified suppliers on the left side of the table, for less complex rigid boards; and fewer in the right-hand columns, for more complex flex boards.

## SUPPLIERS TO THE PrCB INDUSTRY

The manufacturing and assembly of printed circuit boards make up only one part of the PrCB industry. The suppliers to the manufacturers are also critical, spanning a wide cross section of industries. Some of these are specific to the PrCB industry, but most also supply other, related manufacturers.

The basic building blocks of the PrCB come from the glass suppliers, organic resin suppliers, and metals suppliers. Several large multinational companies, including Owens Corning, Ciba Specialty Chemicals, the Shell Group, Dow Chemical Company, BASF, Grupo Mexico, and the Engelhard Corporation, are all in business for the long term and supply many industries in addition to the PrCB industry. There is little likelihood that these companies will cease to make glass fiber or epoxy resin or stop mining copper and refining gold.

The next tier of suppliers is the specialty manufacturers, a potentially weaker link in the supply chain. These companies include Gould Electronics, Inc.; Park Electrochemical Corp.; Polyclad Laminates, Inc.; Isola Group; Taiyo America, Inc.; Rohm and Haas Company; McDermid Corporations; Cookson Electronics; E. I. du Pont de Nemours and Company; Electrochemicals, Inc.; Olec Corp.; Chemcut Corp.; and Technic, Inc. They are responsible for taking the basic raw materials and manufacturing a value-added specialty product that will be used by PrCB manufacturers to build printed circuit boards. Many of the suppliers mentioned above are U.S.-based companies with global operations. They supply materials such as laminate, plating chemicals, imaging films, solder resists, and the equipment to use these materials. Like the rest of the electronics industry, these companies have faced a dramatic decline in U.S. production and revenue from 2000 to 2005. The supplier companies to PrCB manufacturing are particularly affected by these trends for the following reasons:

- The product mix in the United States has shifted heavily to high-performance boards. As suppliers of materials, this supply chain derives its revenue from the square feet of board produced rather than from the value of the finished PrCB. From the suppliers' point of view, the loss of capacity in the United States is significantly higher than simply the loss in the dollar value of finished board products. As a result of this shift, the residual U.S.-based workforce at these suppliers has been drastically reduced, by over 75 percent in some cases.
- Because many of these companies are small businesses, many in this sector have failed or merged with others to attempt to stay financially solvent. Mergers and acquisitions of small businesses have not historically been tracked by the federal government, but the potential impact on the defense industrial base is resulting in increased need for attention to this trend.[13]
- Reductions in the supplier base have reduced contributions to both internal and industry-funded research, development, and product engineering. Prior to 2000, most suppliers would spend a minimum of 10 percent of sales on R&D and technical activities; this has now dropped well below 5 percent. Because suppliers are often the source of new products that spark industry innovations, this loss is difficult to gauge.
- The technical service engineering workforce is almost completely diminished at direct, or "Tier I," suppliers to the PrCB industry in the United States. As is true across most of U.S.

---

[13] S. Patrick. 2005. Remarks presented during a panel discussion at conference titled "U.S. Defense Industrial Base: National Security Implications of a Globalized World," June 2, Industrial College of the Armed Forces.

TABLE 2-4  Companies Qualified to Supply U.S. Military Needs Under MIL-PRF-31032

| Rigid Multilayer MIL-PRF-31032/1 | Rigid Single/Double Sided MIL-PRF-31032/2 | Flex MIL-PRF-31032/3 | Rigid Flex MIL-PRF-31032/4 |
| --- | --- | --- | --- |
| Ambitech | Calumet Electronics | Coretec | Colonial Circuits |
| Calumet Electronics | Coretec (2) | Hans Brockstedt GmbH | Coretec |
| Colonial Circuits | Cosmotronic | Lockheed Martin | Cosmotronic |
| Coretec (4) | Diversified Systems | Printed Circuits, Inc. | Hans Brockstedt GmbH |
| Cosmotronic | Dynamic and Proto Circuits | Sovereign Circuits | Lockheed Martin (2) |
| Diversified Systems | Dynamic Details | Strata FLEX Corp. | Printed Circuits, Inc. |
| Dynamic and Proto Circuits | Endicott Interconnect | Titan PCB East | Sovereign Circuits |
| Dynamic Details | Geometric Circuits | Tyco Printed Circuits (3) | Strata FLEX Corp. |
| Endicott Interconnect | Graphic Electronics | | Tyco Printed Circuits (3) |
| Geometric Circuits | Hans Brockstedt GmbH | | |
| Graphic Electronics | Lockheed Martin (2) | | |
| Hans Brockstedt GmbH | Lone Star Circuits | | |
| Lockheed Martin (2) | Micom Corp. | | |
| Lone Star Circuits | PCT Interconnect | | |
| Micom Corp. | Printed Circuits, Inc. | | |
| PCT Interconnect | Sanmina-SCI | | |
| Philway Products | Sovereign Circuits | | |
| Printed Circuits, Inc. | Teredyne, Inc. | | |
| Sanmina-SCI (2) | Titan PCB East | | |
| Sovereign Circuits | Tyco Printed Circuits (3) | | |
| Teredyne Inc. | | | |
| Titan PCB East | | | |
| Tyco Printed Circuits (3) | | | |

NOTE: Numbers in parentheses indicate the number of separate manufacturing facilities.
SOURCE: Qualified Manufacturer's List, Defense Supply Center Columbus (Ohio), last updated May 6, 2004.

manufacturing, secondary, or "Tier II," suppliers have, with few exceptions, passed direct sales and technical support of PrCB products to Tier II distributors, who have little technical background and have difficulty troubleshooting products in any depth.

- The two preceding factors are reflected in the downsizing of the workforce in that the skilled workers are often the first to leave. The loss of capacity for innovation and the ability to compete for state-of-the-art contracts will continue to erode technical competency over time. These resources most likely will not be replenished in the current environment.
- Capital investments, R&D programs, and technical resources are being heavily emphasized in Asia, primarily China, in hopes of gaining market share for manufactured products. As the United States loses market share, Chinese plants hope to gain it ahead of their competition in Japan, Taiwan, and Germany.

All of these factors are accelerating the already-declining U.S. production of PrCBs to a point that it is difficult today to perceive any technical advantage with respect to the manufacture of PrCBs in North America. In light of the very apparent lack of financial advantage to buying PrCBs in the United States, this acceleration could drive the continued demise of U.S. PrCB manufacturing. If the dozen or so large companies left operating were to abandon their manufacturing of PrCBs in the United States, the suppliers would be forced to follow the business to Asia (either partially or completely) to remain profitable.

Many of the smaller companies that relied heavily on PrCB manufacturing for their revenue will not make the transition and will fail. Other, more diverse companies such as Rohm and Haas or Dupont, could simply exit the U.S. market. Previous experience has shown that such companies may no longer offer their products for sale except through a third-party arrangement. This is very undesirable because it precludes support from the technology-owning partner.

## Materials and Chemistry

The basic building blocks of the vast majority (estimated at more than 75 percent) of the government and military PrCBs manufactured today are epoxy or some other organic resin insulator, glass cloth, and copper. These basic commodity products are used in other industries such as the automotive, marine, construction materials, industrial fabrication products, paints, other electronics manufacturing, and other industrial applications. As an example, Owens Corning, the largest producer of flame-retardant woven glass cloth used to manufacture PrCB laminate, accounts for 80 percent of its revenue in the sales of glass cloth and related products outside the PrCB industry, primarily to the construction materials industry.

The global shift of PrCB production volumes has resulted in numerous new competitors launching products for sale globally from offshore manufacturing locations. In addition, solvent-based resin manufacturing companies have felt pressure to move to offshore manufacturing locations with less demanding disposal regulations. This trend is expected to continue. As a net result, over the next 5 to 10 years the bulk of the base materials for PrCBs may no longer be manufactured in the United States.

The Tier II suppliers, which take the basic building blocks and create laminate, specialty chemicals, and the other ingredients and tools needed to support the PrCB industry, do not generally have the depth of vertical integration that commodities suppliers do—in part because other industries that may be served by these Tier II suppliers are also migrating offshore. The automotive industry, for example, utilizes many building blocks similar to those needed for PrCBs and has become well established offshore. The semiconductor industry, which also shares many Tier II suppliers with the PrCB industry, is no longer dominated by U.S. manufacturing companies. Finally, many products manufactured for the PrCB industry are specific to the industry. While the specifications and requirements may be similar, they are not identical, and therefore these products are not easily cross-marketed.

The simultaneous migration of all industries that buy from the same base could result in a destabilization of these Tier II operations based in the United States. This trend could then spiral, resulting in less technical support, then fewer suppliers, then fewer manufacturers, and so on. Factors that would contribute to this spiral could include increased scrutiny on corporate financial governance, fewer partnership opportunities for R&D programs and joint development efforts, higher costs for business services, and fewer skilled workers.

The most likely short-term outcome under the weight of these compounding pressures is the migration of manufacturing outside the United States.[14] The collapse or complete relocation of these Tier II suppliers to locations outside North America could be one potential outcome that would end the spiral.

## Equipment

The types of equipment used to manufacture government and military PrCBs for legacy, present, and future requirements are particular to the PrCB industry. The companies that produce drill machines, lamination presses, imaging equipment, plating equipment and other finishing tool sets, testing equipment, and routers manufacture them specifically for the PrCB industry. These equipment sets are highly specialized and cannot be used for any other application.

In the past 5 years, many of the U.S. manufacturers of equipment for the PrCB industry have gone out of business, merged with other companies, or followed the supply chain overseas. The closure of so many companies resulted in a glut of used equipment in the United States that could be purchased for 5 to 10 percent of its original value. Over this period, new equipment sales in the United States were very

---

[14] National Research Council. 2004. New Directions in Manufacturing: Report of a Workshop. Washington, D.C.: The National Academies Press.

limited. U.S.-based equipment manufacturers looked to Asia, but found it difficult to compete where local manufacturers quickly reverse-engineer equipment for less than half the cost required for manufacturing in the United States. This trend has spiraled; today U.S.-manufactured equipment has become increasingly expensive in comparison with that available overseas, and companies are struggling.

## BUSINESS CLIMATE FOR PRINTED CIRCUIT TECHNOLOGY MANUFACTURING

The U.S. manufacturing sector faces a number of central challenges, and each has specific relevance to the specialized production of interconnection technologies. The challenges discussed below are part of the changing landscape surrounding manufacturing, defense manufacturing, national security, and economic stability. Both small and large PrCB manufacturers must operate in the current business and industrial climate. Regulations and other constraints—including trade restrictions, environmental regulations, hazardous substance restrictions, labor availability and costs, and insurance and liability costs—influence their operation, both in the United States and globally.

### Cost of Compliance with Regulations

Regulatory initiatives are emerging that require the electronics industry to incorporate environmental, health, and safety considerations into design and manufacturing decisions. Moreover, regulations governing the use, storage, transportation, and disposal of hazardous materials are beginning to influence the electronics manufacturing process. It is hoped that by addressing environmental management issues, electronics manufacturers can reduce both hazardous materials and the generation of hazardous waste. This effort might also lead to improvements in operating efficiencies, reducing procurement costs of raw materials.

The electronics industry is preparing to comply with a number of restricted-materials laws. In 2003, the European Union (EU) enacted the restriction of hazardous substances (RoHS) directive, which bans the use of lead, mercury, cadmium, hexavalent chromium, and certain brominated flame retardants (BFRs) in most electronics products sold in the EU market beginning July 1, 2006.[15] Both business-to-business and consumer products are covered. Although there are some exemptions to the directive's chemical restrictions,[16] by banning the use of critical materials in electronics products sold in key world markets, this directive may result in a significant change in the way products are designed for global sale.[17]

The European Parliament and the European Council are also considering legislation—Regulation, Evaluation, and Authorization of Chemicals (REACH)—that will require industry to prove that chemicals being sold and produced in the European Union are safe to use or handle. REACH policy will require the registration of all substances that are produced or imported into the European Union. The amount of information required for registration will be proportional to the health risks related to the chemical and its production volumes. Companies will also need to seek authorization to sell and produce problematic chemicals, such as carcinogens, mutagens, and teratogens. Toxic chemicals that persist in the environment or that bioaccumulate will also need authorization. The policy is slated for enactment in 2006.[18]

---

[15] European Union. Directive 2002/95/EC of the European Parliament and of the Council of 27 January 2003 on the Restriction of the Use of Certain Hazardous Substances in Electrical and Electronic Equipment (RoHS). Available at http://europa.eu.int/eur-lex/pri/en/oj/dat/2003/l_037/l_03720030213en00190023.pdf. Accessed September 2005.

[16] M. Pecht, Y. Fukuda, and S. Rajagopal. 2004. The impact of lead-free legislation exemptions on the electronics industry. IEEE Transactions on Electronic Packaging Manufacturing 27:221-232.

[17] Note that this is part of a growing global strategy that is supported by a number of environmental organizations and governments. It affects all aspects of lead production, use, and disposal, and has the ultimate goal of separating lead from people wherever possible in accordance with precautionary principles. Appendix E in this report describes the rationale and existing time line for this strategy.

[18] P.D. Thacker. 2005. U.S. companies get nervous about EU's REACH. Environmental Science and Technology Online, January 5. Available at http://pubs.acs.org/subscribe/journals/esthag-w/2005/jan/policy/pt_nervous.html. Accessed September 2005.

California recently enacted the first law in the United States to establish a funding mechanism for the collection and recycling of computer monitors, laptop computers, and most television sets sold in the state. That law, the Electronic Waste Recycling Act of 2003 (SB20), also contains a provision that prohibits a covered electronics device from being sold or offered for sale in California if the device is prohibited from being sold in the European Union by the RoHS directive.[19]

The electronics industry is likewise beginning to take responsibility for its products at the end of their useful life. This responsibility also forms the basis for the "take-back" legislation that is being implemented in the European Union under the Waste Electrical and Electronic Equipment (WEEE) directive, beginning in August 2005.[20] The directive encourages the design and production of electronics equipment to take into account and facilitate dismantling and recovery, in particular the reuse and recycling of electronics equipment, components, and materials necessary to protect human health and the environment.

In the European Union, since July 1, 2003, materials and components have not been allowed deliberately to contain lead, mercury, cadmium, or hexavalent chromium.[21] In addition, strict regulations have been put in place to dispose of components containing lead at their end of life.[22] Lead was classified as category 1, toxic to reproduction (embryotoxic), and as a precaution, the European Union classified lead chromate pigments as category 3 carcinogens.

In the United States, environmental regulation is not moving in the same direction as in Europe. In 2003, the Environmental Protection Agency (EPA) proposed revisions to the definition of solid waste that would exclude certain hazardous waste from restrictions legislated by the Resource Conservation and Recovery Act (RCRA) of 1976 if the waste is reused in a continuous industrial process in the same generating industry. The proposal may eventually exempt all "legitimately" recycled materials from RCRA hazardous-waste regulations. Final action on the proposal is expected in 2006. The EPA is also considering a rule that would exempt electroplating sludge from RCRA hazardous-waste regulations if it is recycled.

In order to ensure that its domestic electronics producers can sell products in the EU market, China has advanced its own RoHS-type law. The draft Management Methods for Pollution Prevention and Control in the Production of Electronic Information Products of the Chinese Ministry of Information Industry would ban the use of lead, mercury, cadmium, hexavalent chromium, and certain brominated flame retardants in consumer electronics and electrical equipment sold in China. South Korea is also considering the enactment of an RoHS-type law, although details are unclear at this time. Asia has many industrial regulations that are not enforced, and considerable time may elapse before attempts at enforcing these regulations are instituted.

While metals are historically important, many electronics products contain brominated flame retardants. Following recent EU moves to ban the use of some brominated flame retardants found to be persistent, bioaccumulative, and carcinogenic, a number of U.S. states have enacted legislation that bans the use of BFRs in consumer goods. Some legislation may include tetrabromobisphenol-A (TBBPA), the leading flame retardant used in circuit boards and computer chip casings. Plastic components of electronics products, such as circuit board laminate, cases, cables, and other structural elements, are likely to be constructed with brominated plastics. There is additional concern over the use of brominated materials owing to their potential to generate halogenated dioxins and furans during open burning and improper incineration.

The cost of compliance with workforce regulations on environment, safety, and health issues can constitute a large part of corporate expenses. As the production of electronics becomes a global enterprise, some of the differences in regulations from country to country may matter less and others may

---

[19] California Department of Toxic Substances Control. Electronic Waste Recycling Act of 2003 (SB20). Available at http://www.dtsc.ca.gov/HazardousWaste/CRTs/SB20.html. Accessed September 2005.

[20] European Union. Directive 2002/96/EC of the European Parliament and of the Council of 27 January 2003 on Waste Electrical and Electronic Equipment (WEEE). Available at http://europa.eu.int/eur-lex/pri/en/oj/dat/2003/l_037/l_03720030213en00240038.pdf. Accessed September 2005.

[21] European Union. Directive 67/548/EEC on the Classification, Packaging and Labelling of Dangerous Substances, Annex 1, as last amended by Directive 2003/32/EC (28th ATP). Available at http://europa.eu.int/eur-lex/pri/en/oj/dat/2003/l_105/l_10520030426en00180023.pdf. Accessed September 2005.

[22] European Union. Directive 2000/53/EC of the European Parliament and of Council of 18 September 2000. End-of-life Vehicles. Available at http://dkc3.digikey.com/PDF/Marketing/ELVdirective_2000-53-EC.pdf. Accessed September 2005.

matter more. Costs related to compliance with regulations include minimizing litigation costs as well as the cost of maintaining balance in the media through public relations. The cost of compliance with regulations may continue to differ substantially in the United States from what it is in Asia, Europe, or the rest of the world. In some cases, however, the global nature of supply and demand can cause regional regulations to become de facto global regulations. Because many companies supply components worldwide, they are finding the cost of producing two types of printed circuit boards, both with and without lead, to be a poor business proposition.

## Challenges in Supply-Chain Management

Outsourcing and offshoring are growing trends with dramatic effects on supply chains and supply-chain management. Within these trends, supply chains are also evolving. With all of these changes, some of the differences discussed here between commercial and military acquisitions and commercial and military supply chains are growing, whereas other differences may be disappearing.

Traditional military procurement has meant onerous accounting processes under the Defense Contract Administration Agency and adherence to rigorous military specifications. Under defense specifications and qualified supplier guidance, companies that wished to supply the government followed a very strict set of rules; the process guaranteed that the product was exactly what the user specified. When the Department of Defense (DoD) moved to performance-based contracting, procurement officials assumed that the product would be produced to the same level of quality as under previous procurements, and that the component performance would be met via that agreement.

During the government's struggles to modernize acquisition, the commercial world has evolved as well. Information technology has had an overwhelming effect on supply-chain management. According to a recent NRC report:

> Information, data communication, and data processing technologies are powerful tools that can be used in every element of the manufacturing enterprise, including just-in-time delivery of raw materials; activities on the factory floor; shipping; marketing; and strategic planning. These tools can manipulate, organize, transmit, and store different types of information in digital form. The impact of these technologies has been compared to that of the technological advances that spurred the Industrial Revolution.[23]

Some industry analysts estimate that total supply-chain management, including remote step-by-step process controls, will be ubiquitous within 5 years. A driver for this potentially disruptive change in PrCB manufacturing is the introduction of RoHS. To guarantee, for example, the absence of lead in a board, a manufacturer may need to produce the same level of documentation once required by DoD. This is proving to be especially necessary when a product changes hands more than once during manufacturing. Advances in systems, processes, documentation, and information technology may help to make this kind of tracking inexpensive for all components.

In defense acquisition, the transition to performance-based contracting has been difficult. Expectations exist for materiel to be supplied, but there is little management of the supply chain. In the commercial world, supply-chain management involves understanding the daily status of materials sources, knowing the potential risks of different suppliers, and making sure that multiple sources are available. It also involves effective communication of projected needs and time lines to the suppliers. While government purchasers are able to direct integrators to buy from particular sources, such as a directed buy for "critical" components, this is not normally done as part of a strategy to ensure constant supply. In the commercial world, these practices are common and are driven by the bottom line; good companies are always assessing their supply-chain risk.[24]

DoD (and the rest of the federal government) has never needed to track its supply chain in the same way that a company would. It is becoming clear, however, that some oversight and assessment of supply-chain capabilities are needed. It is likely that the solutions that have been developed by responsible companies with similar production volumes and applications—for example, biomedical

---

[23] National Research Council. 2004. New Directions in Manufacturing: Report of a Workshop. Washington, D.C.: The National Academies Press, p. 14.

[24] S. Cohen and J. Roussel. 2004. Strategic Supply Chain Management. New York: McGraw-Hill.

devices or commercial aerospace components—may be the most useful models for future supply-chain management.

According to Steven Mather of the Computer Sciences Corporation, "Ultimately, force readiness is a primary concern of the Army, whereas in the commercial sector the primary concern is profitability."[25] DoD acquisition managers who are attempting to integrate commercial purchases into their portfolios are also beginning to incorporate the responsibilities of being a good customer.

## Cost of a Skilled Workforce

Traditionally, the cost of workers in a manufacturing enterprise had been limited to wages and benefits. In recent years, the differential among global wages has received most of the attention from company managers. Benefits are also an important factor—global differences in the cost of health insurance and pensions are becoming a larger issue than wage differentials in many cases.

A number of other costs are unaccounted for in most workforce equations. The cost of maintaining workforce skills in a changing industry can be very high. Hidden costs also exist in maintaining a corporate history in the art of manufacturing processes for some defense components in legacy systems. For a small shop, the loss of the skills and background knowledge of one or two people can be disastrous.

In addition to such changes in workforce accounting, organizations are being pressured to provide enhanced levels of service to their employees. Doing so is key to retaining high performers, avoiding the costs of excessive recruiting and training, and ensuring that workers are prepared to succeed in a changing technology environment. Many manufacturing companies are facing seemingly contradictory goals, needing both to cut workforce costs and at the same time to invest in the workforce so that it can do more.

## Challenges in Innovation

In electronics as in all technologies, product cycles are getting faster and technology complexity is increasing.[26] This cost of keeping pace with all of this new technology is increasing in parallel. In addition to an increasing demand for new products, the demand for traditional products with innovative features is also increasing. To remain competitive, engineers are seeking ways to add more capabilities and compatibilities to all products.

In many cases, accomplishing and even implementing a technology can come well before scientific understanding of the basic underlying principles is achieved. To understand many new technologies, a more interdisciplinary approach and more innovative tools are needed. Complex multilayered PrCBs are stressing the state of current knowledge and will require ever more know-how and scientific investigation. While it is still true that fundamental understanding may be necessary for optimizing or adding capabilities, it is important to realize that we may not truly understand the technology we use today. This can mean that new technologies cannot be fully exploited without investment and ideas.

In their early development, PrCBs were relatively simple structures. The desire for ever-greater system performance has driven the PrCB industry to combine disciplines, technologies, and tools in order to achieve tremendous complexities; this process calls for ever-increasing knowledge of materials and processes. Today, the most advanced interconnection technologies, including the creation of metal/plastic composites and many-layered structures, and the combining of optical and electronic phenomena, are seeking to exploit phenomena that are beyond known and tested practices. Various factors—the variety of raw materials, the decreasing thickness and size, and the challenges of heat dissipation—are all stressing the current knowledge of what can be done and how to do it.

---

[25] P.E. Clarke. 2003. Re-engineering the Supply Chain. Military Information Technology 7. Available at http://www.military-information-technology.com/article.cfm?DocID=28. Accessed September 2005.

[26] Note that the product cycle of a PrCB is generally bound to that of the product it serves. Product cycles for items that incorporate PrCBs range from as short as 6 months for a cellular telephone, to up to 5 years for automotive components, and as long as 30 years for infrastructure applications.

While innovation occurs everywhere, not just in research laboratories, access to the necessary know-how, research equipment, or even workers with scientific or engineering training is something that small or niche manufacturers cannot typically afford. Only enterprises with sufficient scale and scope— typically larger companies and government agencies—can pay for this type of knowledge generation.

One of the biggest challenges for innovation facing the specialty PrCB industry is the difficulty of efficient low-volume operation. Whereas the typical production of microchips may be millions per day, the many configurations of boards mean much smaller quantities of each. DoD routinely orders as few as one or two replacement parts. Increasingly, manufacturing demands six sigma and higher quality;[27] it is impossible to even gather those statistics in these low-volume production rates. Therefore, reliability must be engineered in a different way and will require new levels of innovation.

Efficient low-volume specialty production could become an alternative paradigm that could offer a competitive advantage to innovative partners in this industry. It is important to note that if such processes were to become practiced worldwide, this approach would offer DoD the ability to produce the needed specialty parts in desired locations with low investment.

Another major driver requiring innovation is that the chemical processes for manufacturing PrCBs are some of the most environmentally difficult. Waste-disposal costs are very high, and closed-system processes are needed. In addition, the complex chemistry means that processes can be easily upset and one imbalance can result in an entire manufacturing run needing to be scrapped. In many cases, the failure is not known until the boards are tested.

A final note on the need for innovation is the coming and overwhelming challenge to manufacture all electronics without lead solders, lead-based ceramics, or lead coatings. Though the engineering challenges are proving to be problematic on many levels, the ability to seamlessly integrate no-lead technology into current and legacy systems may be even more difficult.

## KEY FINDINGS AND CONCLUSIONS

By a number of measures, the PrCB industry in the United States is in a steep decline. Changes in the number, size, and scope of the companies that manufacture PrCBs appear to be a result of the evolution of global markets and production. The companies that supply the PrCB manufacturing industry are particularly affected by these trends. The committee finds that the 400 or so U.S. companies may not be able to stay competitive in this high-technology area. It is probable that without outside support, these small PrCB suppliers will not continue to be able to meet the requirements of U.S.-manufactured PrCBs for government and military applications.

Interconnection technology—boards and other printed circuitry—is a key element of commercial and defense systems. For the companies that meet U.S. military needs today to sustain their performance over the long run they will need a direct linkage to the technology advancements of the global PrCB industry. It is becoming apparent that DoD purchases from military suppliers will not be large enough to create that linkage. Therefore, the loss of this industry in the United States may adversely affect the ability of the remaining companies to supply future military needs.

---

[27] Six sigma is a data-driven approach and methodology for eliminating defects. It is intended to achieve six standard deviations between the mean and the nearest specification limit for any production process.

# 3

# Military Needs for Printed Circuit Technology

Computers and electronics are estimated to account for more than one-third of defense procurement spending, and this proportion is steadily increasing.[1] Interconnection technologies in the form of printed circuit boards (PrCBs) are integral to all of these systems. Printed circuit boards are fundamental to the operation of military navigation, guidance and control, electronic warfare, missiles, and surveillance and communications equipment. High-density, highly ruggedized, highly reliable interconnection technology is essential to the implementation of much of this country's superior weaponry. The importance of PrCBs to all military missions—legacy, current, and future—cannot be overstated.

However, as with many other industries, the acquisition of PrCBs poses a predicament to Department of Defense (DoD) purchasing. The difficulty lies in the DoD's unique requirements, its diminishing purchasing position within the overall market, and its ever-increasing demand for higher technical performance at affordable cost. Industry investments by PrCB producers in both manufacturing equipment and manufacturing expertise are focusing on the high-volume, low-cost growth segment rather than on the high performance, reliability, and extreme environmental tolerances required for DoD products.

## DEFENSE REQUIREMENTS

Goods manufactured for defense are very different from goods manufactured for commercial use in some obvious ways, although many similarities exist as well. In the area of electronics, the similarities may be the most apparent. Military trucks may be very different from commercial trucks of similar size, but the PrCB that controls the fuel injectors in both types of vehicles may be exactly the same. In another design scenario, however, they may be wholly different. Figure 3-1 shows one framework for these varying requirements.

In terms of defense priorities, DoD must first respond to DoD technology and product development requirements; second, DoD logistics must meet delivery requirements during peacetime and/or periods of conflict or international tension; and finally, all activities within these steps must preclude unauthorized transfer of technical information, technologies, or products within the United States or to third parties.[2] These three points correspond to broad principles of strategic alignment, assured supply, and industrial security.

---

[1] W. Andrews. 2003. Computers and electronics component of DoD budget continues to rise. COTS Journal 5:54-58.

[2] Department of Defense. 2004. Defense Industrial Base Capability Study: Battlespace Awareness. Washington, D.C.: January, p. ix.

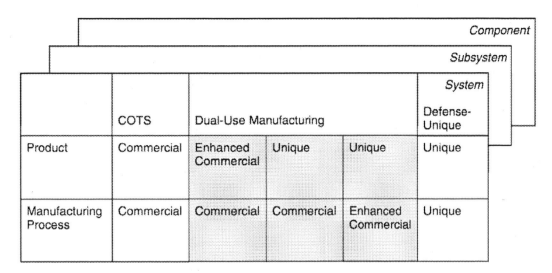

| | COTS | Dual-Use Manufacturing | | | Defense-Unique |
|---|---|---|---|---|---|
| Product | Commercial | Enhanced Commercial | Unique | Unique | Unique |
| Manufacturing Process | Commercial | Commercial | Commercial | Enhanced Commercial | Unique |

FIGURE 3-1 Product and process requirements in a commercial-military integration framework.
NOTE: COTS, commercial off-the-shelf. SOURCE: National Research Council. 2002. Equipping Tomorrow's Military Force: Integration of Commercial and Military Manufacturing in 2010 and Beyond. Washington, D.C.: National Academy Press, p. 2.

## Demands on Technology

Printed circuit technology as designed for military applications tends to be highly specialized, primarily because of the special functions and packaging requirements for DoD weapons systems. Military PrCBs are also produced at low volumes compared to those produced for nonmilitary PrCB applications. For example, one of the highest-volume systems planned for use across the military services is the Joint Tactical Radio System (JTRS), which had been projected to implement up to 30,000 units in the next 2 years, and potentially 108,000 over the next 10 years.[3] Even this capacity is much lower than that for many of the least-popular commercially marketed radios or cellular telephones.

The requirements of electronics for military use also tend to be far more demanding than those of electronics for commercial applications. A commercial component is designed to have a typical lifetime of 2 to 5 years; after that much time in use the technology becomes obsolete, so it is not economical to design such components to survive very much longer. A typical military component can take substantially more time to design, qualify, and implement than is required for commercial components; the expected lifetime for military components is typically more than 5 years and is often extended to 15 years or longer. A prime example of how the military can extend a system's life is described in Box 3-1 for the SLQ-32 Electronic Warfare System.

In addition to having long life, military components are generally expected to be more reliable, robust, and rugged than most commercial products are. Moreover, use conditions are in many cases quite different from those of commercial technologies: military electronic systems are expected to perform in battlefield conditions, with extremes in temperature, humidity, vibration, and impact, in addition to surviving the possibility of salt spray, blowing sand or dust, and solar radiation.

---

[3] Congressional Budget Office. 2003. Appendix A: The army's current communication initiatives. Pp. 31-35 in The Army's Bandwidth Bottleneck. Washington, D.C.

---

**BOX 3-1**
**The SLQ-32 Electronic Warfare System**

The AN/SLQ-32(V) Electronic Warfare System can be found on more than 150 ships of the U.S. Navy as well as on 13 U.S. Coast Guard cutters. It was the second-most-deployed combat weapons system in the U.S. Navy during Operation Iraqi Freedom, with 57 systems deployed from March through April 2003. The SLQ-32 (as it is commonly referenced) is designed to give ships early warning against anti-ship guided missiles (ASGMs) and to provide combat system support. In short, the system equips ships with the capability to defend themselves against ASGM attacks.

The SLQ-32 Program was initiated by the Department of Defense (DoD) in 1972, with the first contract awarded in 1977 and the first product installed in 1979. In 1983, an improvement plan was initiated to enhance the system, but between 1996 and 2002, the funds were phased out and redirected to a newer system. When the development of the latter system was cancelled, the SLQ-32 was reinstated and required to remain tactically viable and supportable until at least 2025. The Navy's Surface Electronic Warfare Improvement Program aims to improve the current SLQ-32 system by spiral development until the system meets the requirements of the original replacement program. While this development is under way, ships continue to rely on current SLQ-32 systems. Some components of the SLQ-32 will need to be supported for at least 20 more years.

Keeping a steady supply for SLQ-32 components can be difficult: 171 assemblies in the SLQ-32 system contain obsolete components, and of these 171, 34 assemblies have less than a 5-year supply. The best approach is to keep suppliers running manufacturing lines that produce these components, but this is a challenge because of the low volumes—sometimes only a few per year are needed—and suppliers prefer to marshal their limited resources toward more lucrative orders.

When no more suppliers are available or willing to bid on an order, the part becomes obsolete. When this situation is well understood, the government can evaluate whether the supplies in the inventory will be sufficient for the rest of service, and if not, can buy larger quantities of the component before manufacturing ends. There are several options when a part becomes truly obsolete: newer or off-the-shelf components can be qualified to replace the obsolete parts or subassemblies; new vendors can be developed (in the United States or abroad) for some products; or, finally, the government can step in and use existing capability to take over the manufacturing and maintenance of these components.

The beam forming lens (BFL) is considered to be the most critical component of the SLQ-32. BFLs are used to achieve a 100 percent probability of signal intercept over a 360° field of view over the full frequency range of the Electronic Warfare System. BFLs are constructed of matched pairs, some of which are among the largest printed circuit boards made. Unfortunately, the original equipment manufacturer (OEM) had become unreliable: the last order that it produced took a year to arrive, and the OEM did not bid on a subsequent order for BFLs.

Looking for other suppliers for BFLs, the government ran into a number of difficulties. Because only the OEM had the board design details, reproducing the board pattern was a challenge. Attempts to scan the pattern at a commercial site were unsuccessful, and prints provided in the drawing package were not adequate to reproduce the patterns. In the end, the government decided to mitigate risks by using in-house production capability to manufacture the component. This process was complicated and costly: it involved lengthy reverse engineering and now utilizes government manufacturing resources. Ultimately, industry support does not last the entire life of many military components, which can be more than 40 years in some cases.

SOURCE: Naval Surface Warfare Center Crane Division.

---

An array of nondefense applications—air-traffic control, postal sorting, and law enforcement uses— have elements similar to those of military components. Because these are also funded with taxpayer dollars, they are equally subject to cost pressures. Operating rates in these applications—whether for defense or nondefense purposes—can vary from the most demanding daily use to sitting at-the-ready for years between on-off cycles—both states involving rapid environmental changes. In addition to such requirements, military personnel routinely expect systems to operate outside the strict performance parameters for which they were designed. Finally, with the growing degree of asymmetrical threats, the level of performance applied to electronics used in combat-ready equipment is also expected for tactical and support equipment. In short, defense requirements exceed commercial design standards in almost every category.

## Demands on Supply Chains

The challenge to the military supply chain to ensure supply is further complicated because it must not only provide for current military needs, but it must also prepare for the far future while maintaining in readiness much equipment that was deployed in the distant past. A further complexity is that the supply chain must also operate laterally—supporting equipment in locations, conditions, and time frames different from those for which the equipment was originally designed.

The military needs to service and maintain equipment no matter when or where it is used. Many systems go through a number of application "lives" in various training or conflict situations or those requiring a military presence. As equipment is transferred from one soldier or unit to another, the military retains responsibility for it. Also, when a piece of equipment finally outlives its usefulness, the military pulls it back through a reverse supply chain and disposes of the item.

DoD must decide whether to "make, buy, or fix" components and systems as a matter of course throughout its supply chain. However, many times this decision step is not adequately factored in to current practices. The military services tend to maintain equipment no matter when a product was originally manufactured, in part because fielded equipment becomes the operating unit's problem; it is no longer the producer's responsibility. Many times, a particular PrCB part is no longer manufactured, and in some instances the original drawings and manufacturing specifications are not available. The ability to repair, replace, or reengineer legacy components is a valuable aspect of the DoD supply chain, but it is also very costly. Some OEM and aftermarket suppliers command very high prices for parts that are difficult to make with newer equipment, or parts for which the older design drawings are unavailable.

Different levels of technology needs are described according to time frames from past/legacy to far future in Table 3-1. Each time frame has different sources and requirements. For example, some older legacy PrCB components will have their own specific drawings and specifications; some will fall under MIL-PRF-55110; some will fall under the newer preference, MIL-PRF-31032; and still others will be specified only as to their performance and are intended to be purchased as commercial off-the-shelf (COTS), or may be specified to be made as per IPC-6012 Class 3. Ultimately, however, each decision point to determine the links and branches of the supply chain must result in good value to the warfighter. Decisions based on scenarios and consequences need to seek out robust rather than optimal strategies, and they need to employ adaptive strategies that can evolve over time in response to new information.[4]

Because of all of these constraints, as well as procurement processes, cycle time, qualification procedures, cost of shelf life, and a variety of other factors, DoD electronics tend to be more than one or two generations behind commercial technology. Experience has shown that commercial technology cannot be directly inserted for defense technology, but usually needs to be transitioned with thought and perhaps with constraints. For example, in cases where commercial technology can be directly inserted into military systems, the design goals, materials, and manufacturing processes should be known. In certain cases, some features required in government applications are so unusual that commercial technology cannot be modified but must be redesigned. To supply the warfighter effectively, government officials need to understand the manufacturing technology behind the needed component. This understanding implies a dual-use strategy rather than a pure COTS mode, in which components can be procured based on a strong dual (commercial and defense) industrial base.

## Demands on Assurance

The Department of Defense has strong concerns about the unauthorized transfer of critical technical information, technologies, or products within the nation or to third parties. The problem is which

---

[4]  R.J. Lempert, S. Popper, and S.C. Bankes. 2003. *Shaping the Next One Hundred Years: New Methods for Quantitative, Long-Term Policy Analysis.* Santa Monica, Calif.: RAND Corporation.

TABLE 3-1 Technology Assessment for Different Military System Time Frames

|  | Past/Legacy | Current | Near Future | Far Future |
|---|---|---|---|---|
| When system was specified or purchased | More than 10 years ago | Fewer than 5 years to present | Present to more than 2 years | More than 5 years from now |
| Level of technology | Obsolete technologies with diminished manufacturing sources | Existing technology | Today's known and some cutting-edge technology | Not yet known |
| Source of replacement (or repair) parts | Combination of small shops and organic facilities | Replacement parts bought when systems were procured | Not yet known | Not yet known |
| Critical factor for success | Part repair or replacement (including reengineering) | Cost-effective replacement | Design for support in today's acquisition environment | Design for support in a new acquisition environment |
| Potential consequence of failure | System failure, leading to new system design/acquisition | High cost limits other needed acquisitions | Continuance of today's supply-chain challenges | Inability to take advantage of new technologies |

technical information, technologies, or products are critical, because the cost penalties, in terms of operational restrictions, for protection against unauthorized transfer can be very high. It is for this reason that DoD minimizes the use of security classification as a means of protecting its information, capabilities, and technologies.

While many industries and companies are very protective of their technologies and trade secrets, many aspects of defense systems must be further protected under penalty of law. Therefore, reliance on third parties (other than the government and primary contractor) to manufacture some or all components or systems used by the warfighter can provide a variety of challenges and burdens on government acquisition practices. To achieve a high degree of assurance in a particular technology creates a major burden and challenge for the government owing to these constraints.

Attempts to limit defense manufacturing to operations within the United States have resulted in charges of protectionism. These attempts have not been supported by the current administration or the Congress. The chairman of the Federal Reserve Board, Alan Greenspan, states that "protectionism will do little to create jobs; and if foreigners retaliate, we will surely lose jobs. We need instead to discover the means to enhance the skills of our workforce and to further open markets here and abroad to allow our workers to compete effectively in the global marketplace."[5]

The military thus is faced with a challenge: that of buying components in the global marketplace, while protecting technical specifications of those components from some of the participants in the same global marketplace. To address this dilemma, the United States has a number of controls on the export of technology that affect a number of existing and emerging technologies. While such controls are not currently mirrored worldwide, there is a growing trend toward the use of export and import controls by a growing number of countries.

In the United States, the Department of State's Office of Defense Trade Controls and the International Traffic in Arms Regulations (ITAR) control the permanent and temporary export and temporary import of defense articles and defense services. This control is exerted primarily by the taking

---

[5] Federal Reserve Board. 2004. Remarks by Chairman Alan Greenspan on the Critical Role of Education in the Nation's Economy, at the Greater Omaha Chamber of Commerce 2004 Annual Meeting, Omaha, Nebraska, February 20.

of final action on license applications and other requests for approval for defense trade exports and retransfers, and by handling matters related to defense trade compliance, enforcement, and reporting. A major element contained in the ITAR is the U.S. Munitions List (USML), the official list of the types of items controlled by the Department of State.

The Export Administration Regulations (EAR) are issued by the Department of Commerce's Bureau of Export Administration. The export control provisions of the EAR are intended to serve the interests of the United States in the areas of national security, foreign policy, nonproliferation, and short supply, and, in some cases, to carry out its international obligations. Some controls are designed to restrict access to dual-use items by countries or persons that might apply such items to uses opposed to U.S. interests. These provisions include controls designed to stem the proliferation of weapons of mass destruction and controls designed to limit certain countries' military capability and ability to support terrorism. The EAR also include some export controls to protect the United States from the adverse impact of the unrestricted export of commodities in short supply.

The Militarily Critical Technologies List (MCTL) serves as a technical reference for licensing and export control by the U.S. Customs and Border Protection Bureau and the Departments of State, Defense, Commerce, and Energy. The MCTL is a compendium of existing goods and technologies that DoD assesses would permit significant advances in the development, production, and use of military capabilities of potential adversaries. Decisions about candidate technologies are made by technology working groups, composed of experts from business, government, and academia. While the MCTL provides guidance in the development of export control regulations, it is not authoritative. Items may be in the MCTL that are not export controlled and vice versa.

The MCTL has two parts—the MCTL itself and the Developing Science and Technologies List (DSTL). The DSTL is a compendium of scientific and technological capabilities being developed worldwide that have the potential to enhance or degrade U.S. military capabilities significantly in the future. It includes basic research, applied research, and advanced technology development. Electronics technology is discussed in Section 8 of the DSTL, which was most recently updated in 2000. The lag of this update with respect to the ever-faster increases in new technology is significant. In addition, it is significant that neither the MCTL nor the DSTL covers manufacturing innovation explicitly.[6]

Export controls are not designed to meet the U.S. defense need for assured supply. They have little to do with providing assurance with respect to actual products or technologies that DoD would acquire; rather, their purpose is to keep some level of weapons technologies from potential adversaries.

## DEFENSE MANUFACTURING ENVIRONMENT FOR PRINTED CIRCUIT TECHNOLOGY

The transformation of the military in the post-Cold War era has affected the defense industrial base in a number of ways. New demands on the rate of technology change have driven many of the changes in defense acquisition.[7] Military leaders see a need to respond to a wholly different array of threats in the future, which has led to a focus on the transformation of the entire defense enterprise. According to Secretary of Defense Donald H. Rumsfeld, "The Department is in need of change and adjustment. The current arrangements, designed for the Cold War, must give way to the new demands of the war against extremism and other evolving challenges in the world. We face an enemy that is dispersed throughout the world. . . . Our enemy is constantly adapting and so must we."[8]

A universal view of what constitutes the defense industrial base is difficult to pin down. This base evolved through the 1950s and 1960s into an assortment of companies that came to depend on the defense establishment as their primary and sometimes only customer, to sustain them. In turn, the military depended on these companies for a global technical advantage. Within this paradigm, defense spending drove the cutting edge of new technology, and these companies then spun some of these innovations out into commercial applications. Such innovations as computers, digital cameras, and hand-

---

[6] The MCTL and DSTL can be accessed at http://www.dtic.mil/mctl/. Accessed October 2005.

[7] Office of the Deputy Under Secretary of Defense (Industrial Policy). 2003. Transforming the Defense Industrial Base: A Roadmap. February. Available at http://www.acq.osd.mil/ip. Accessed September 2005.

[8] Secretary of Defense Donald H. Rumsfeld. Remarks at the Base Realignment and Closure (BRAC) Commission Hearing, May 16, 2005. Available at http://www.brac.gov/docs/BRACHearingFullTranscript16MayPM.pdf. Accessed October 2005.

held Global Positioning System units are all commercial technologies with strong roots in defense spending, reflecting the luxury that the nation once had of drawing on an indigenous and dominant technology innovation base.

Over time, beginning in the 1970s and continuing in the 1990s, all of this changed. Commercial spending on research and development (R&D) outstripped military spending, and technologies began to be spun into defense applications. Defense technology once came directly from military-focused R&D; it is estimated that during the Cold War, defense spending funded the vast majority of R&D in the United States. However, since the mid-1960s, the percentage has fallen to slightly more than half,[9] and the overall R&D spending as a percentage of gross domestic product (GDP) has dropped from a high of nearly 3 percent to less than 1 percent.[10] At the same time, graduate students became dominantly non-U.S. citizens. The paradigm is continuing to shift today as global universities are producing higher-quality graduates and the knowledge base is becoming truly global.[11]

As these changes have occurred, the traditional underpinning of defense spending for the high-technology economy in the United States has shifted as well. The interdependence of a strong national defense capability and a strong economy were axiomatic throughout the post-World War II era in any number of ways. One key differentiating element today is the growth in complexity that has been enabled by electronics. In the old paradigm, the sheer volume and mass of equipment in the warfighter's control ensured dominance. In the new paradigm, the precision afforded by embedded electronics can provide the same dominance with much lower raw-material and manufacturing costs. And in this new paradigm, commercial technology dominates.

As a result, DoD initiated a number of policies to draw on commercial successes to insert new technologies into defense systems.[12] These have not been sustained, however, through various defense administrations, and it is apparent from its acquisition decisions that DoD has never considered that supporting the U.S. economy is its responsibility.[13] Yet the interdependence of the defense and commercial industrial base can be asserted; key factors include the importance of sustaining technological superiority, a declining production base for new weapons, and advantageous trends in the commercial manufacturing sector.

Much of the post-Cold War policy for defense acquisition has been targeted to encourage this kind of commercial-military integration. The transformation strategy implicitly relies on the rapid introduction of new technology and rapid industrial response for the replenishment of weapons, spare parts, and other consumables essential to readiness and sustainability. Transformation also has embraced a new drive for unmanned and remote systems, which need to be reliable with a minimum of hands-on maintenance.

Today, however, this integration is difficult to measure. A recent National Research Council report finds that "our military decision makers today are trapped in an acquisition strategy that depends on an industrial base that cannot respond quickly enough to meet the demand for new and modified military systems expected to result from the stepped-up tempo of future military operations."[14]

Today, a handful of U.S. companies make up the defense integrated products industry and remain funded primarily by defense spending. The factors contributing to this consolidation include increased integration of commercial and military manufacturing and the escalating technological intensity of defense materiel.[15] These companies produce "defense-unique" products for which the military is the primary customer. These products include tanks and armored vehicles, ammunition and ordnance, aerospace vehicles and ships, and a variety of electronics for search and navigation electronics, night vision, and other specialized applications.

---

[9] American Association for the Advancement of Science. 2005. Federal Spending on Defense and Nondefense R&D. Available at http://www.aaas.org/spp/rd/histde06.pdf. Accessed September 2005.

[10] American Association for the Advancement of Science. 2005. U.S. R&D as Percent of Gross Domestic Product. Available at http://www.aaas.org/spp/rd/usg03.pdf. Accessed September 2005.

[11] R. Van Atta, M. Lippitz, and R. Bovey. 2005. Defense technology management in a global technology environment. IDA P-4017. Alexandria, Va.: Institute for Defense Analyses.

[12] J.S. Gansler. 1995. Defense Conversion: Transforming the Arsenal of Democracy. Cambridge, Mass.: MIT Press.

[13] U.S. Department of Defense. 2003. Transforming the Defense Industrial Base: A Roadmap. Washington, D.C.: DoD Office of Industrial Policy.

[14] National Research Council. 2002. Equipping Tomorrow's Military Force: Integration of Commercial and Military Manufacturing in 2010 and Beyond. Washington, D.C.: National Academy Press, p. 10.

[15] K. Flamm. 2005. Post-Cold War policy and the U.S. defense industrial base. The Bridge 35:5-12.

It is in the supply base for these defense-unique products that commercial-military integration is expected to occur. The fact that potential adversaries have easy access to the same commercial technology provides a compelling additional reason for concern. An excellent case in which to examine supply base issues is that of PrCBs, which are a broadly applicable, ubiquitous technology with equal importance to commercial and military needs.

While the military provided the original testbed for many computers and microelectronics, defense needs are not the driver for the newest technologies in these fields in most cases. Consumer demand for many advanced electronic products has far outstripped the volume required by the military.[16] At the same time, the military depends increasingly on electronics to meet its mission as a smaller, more agile, transformed military force. Today, costs for military electronics are estimated to account for almost 50 percent more than costs for military aircraft—75 billion dollars for electronics versus 50 billion dollars for aircraft.

Electronics is perhaps the easiest of all defense technologies to globalize because of the direct ties to commercial technology. Because of this globalization, the connection between technology investment in direct support of defense capabilities and economic strength has been severed. This severing was intentional, because as a nation the United States does not want to limit commercial production and sales to the United States. This would make no economic sense in today's global markets. However, it does not follow that DoD cannot help this nation be a strong global competitor.

It has become clear that the DoD's technology investment is more closely tied to commercial applications than ever before. These new commercial technologies can provide worldwide benefits in electronics, transportation, medicine, and energy, to name a few examples. Because the benefits extend beyond defense, however, does not mean that DoD should stop investing. Instead, DoD needs to find ways to invest that will allow it to best capture the benefit. Admittedly, this is not a trivial problem.

## Scoping the Challenge

The Department of Defense purchases items with more than 8 million different part numbers, or National Stock Numbers (NSNs).[17] The Defense Logistics Agency (DLA) provides supply support and technical and logistics services to the military services and to several civilian agencies. The DLA manages over 5.2 million items that support individuals and the services' weapons platforms. As part of this mission, it supplies 95 percent of repair parts for critical assets such as aircraft, tanks, and other weapons platforms.

Every day, the DLA receives more than 54,000 requisitions. The agency processes nearly 8,200 contract actions daily and does business with nearly 24,000 different suppliers. More than 21,000 persons are employed by the DLA to carry out this mission, although this total is reduced substantially from 65,000 in 1992, just after the end of the Cold War.[18,19]

For printed circuit boards, the logistics data are mixed. A cursory look at the suppliers of boards[20] to the DoD Defense Logistics Agency yields a list of the following top 11 companies:

- Northrop Grumman Systems Corporation;
- Lockheed Martin Corporation;
- Agilent Technologies, Incorporated;
- Litton Systems, Incorporated;
- Rockwell Collins, Incorporated;

---

[16] K. Flamm. 2005. Post-Cold War policy and the U.S. defense industrial base. The Bridge 35:5-12.

[17] A National Stock Number (NSN) is a 13-digit number assigned to an item of supply. It consists of the 4-digit Federal Supply Class (FSC), and the 9-digit National Item Identification Number (NIIN). The NSN is used as the common denominator to tie together logistics information for an item of supply. An NIIN is a unique code assigned to each item of supply purchased, stocked, or distributed within the federal government; when combined with the FSC, it composes the NSN.

[18] Facts and Figures About the Defense Logistics Agency. Available at http://www.dla.mil/public_info/facts.asp. Accessed September 2005.

[19] K. Horn, C. Wong, E. Axelband, P. Steinberg, and I. Chang. 1999. Maintaining the Army's "Smart Buyer" Capability in a Period of Downsizing. Santa Monica, Calif.: RAND Corporation.

[20] Federal Supply Class 5998: Electronic Assemblies; Boards, Cards, Associated Hardware.

- Smiths Aerospace;
- BAE Systems;
- ITT Industries;
- Honeywell;
- Boeing; and
- Raytheon.

It should be noted that only three of these companies, Lockheed Martin, Litton Systems, and Rockwell Collins, are board manufacturers. The rest purchase boards for their systems. These other eight companies, therefore, outsource the manufacturing of the boards and, given the observed trends, it is certainly possible that many of them source these components worldwide. DoD does not currently track this possibility, nor does it require its contractors to do so.

The inability to trace the sources of potentially critical components, such as PrCBs, is a potential difficulty for DoD. Understanding sources and supply chains for both new and replacement parts is a best manufacturing practice for any large or small organization. It is possible that policy changes (and the resulting decreased workforce in defense acquisition and logistics) have contributed to a diminished requirement to manage the supply chain of these types of components.

It is also interesting to note that most of the companies on the list above are primarily defense suppliers. Post-Cold War acquisition practices have led to a massive consolidation of defense suppliers (both vertically and horizontally). This change, in turn, has created a perceived separation of the defense industrial base from commercial industrial base. This phenomenon is understood to be driven by a number of well-known factors, primarily the burdensome paperwork and separate accounting systems required to satisfy defense contracting, which has resisted the streamlining forced in the commercial sector by competitiveness, and shrinking markets and declining defense budgets. This separation is observed even in companies that supply similar products to military and commercial markets.

An overarching factor has also been the increasing dematerialization of warfare—which is realized by more electronic and smarter systems in use at all levels of combat and correspondingly fewer tons of metal per warfighter. This dematerialization has resulted in a greater dependence of national security efforts on information and information technology and less dependence on material-intensive tanks and planes.

## Risk and Sustainment

A system can fail for want of a structural or supporting component even though the component does not directly enable the sought-after warfighter capability. However, the idea of making timely and detailed assessments of every one of the more than 8 million components that DoD buys is untenable. Current DoD practices require DoD to rely on the market but to watch out for developing problems. Program managers have the freedom to emphasize use of standard components and open system architectures in order to maximize the number of sources available to their programs. However, as issues arise, DoD must have the ability to link the component through the systems to warfighter capabilities and to integrate impacts and remedies into defense decision making.[21]

In current DoD assessments, circuit boards do not appear as a critical technology, nor do they appear as a critical component in the subset of critical technologies that warranted further analyses in the MCTL or DSTL. This is because electronics interconnection technology does not provide a "direct warfighter capability"—the main criterion used to identify critical technologies and components. However, this does not mean that circuit board technology is not important to the defense enterprise. Indeed, it is a prime example of the structural and supporting technologies that must be available to implement critical technologies and to deliver systems to the warfighter. These enabling technologies and components far outnumber currently identified critical technologies and must also be available to DoD when necessary.

The intent of acquisitions decision makers is to assess the risk that supply of materiel brings to mission success. Such risk can be assessed in a reactive or proactive manner. DoD may choose to be more proactive regarding critical technologies and components and to focus its attention on those items

---

[21] Department of Defense. 2004. Defense Industrial Base Capability Study: Battlespace Awareness. Washington, D.C.: Office of the Secretary of Defense, p. 4.

while allowing itself to adopt a reactive posture regarding other important technologies, especially where there are large commercial forces at work. The difference in approach depends on a number of factors.

For example, a number of options exist for improving the reliability of access to PrCB functionality. Some proactive approaches are to stockpile raw materials or to allocate more funds to keep a larger number of manufactured parts in inventory. It is important to realize that some of these options would support innovation and others would not. Buying too far ahead is not a simple option because PrCBs have a solderability shelf life of between 6 months and 2 years. The shelf life is limited by a number of factors, such as humidity, packaging media, and coating composition. DoD may choose, in some cases, to keep populated boards for specific high-value applications, and may also invest in higher-cost storage to extend the shelf life of bare boards.

A potentially costly alternative is to manufacture unneeded parts in order to keep captive manufacturing in operation. This means that highly reliable capabilities will be available when needed. These manufacturing capabilities can be in-house, on military reservations, or in captive small shops. Standing capacity can also be achieved through agreements forged with larger manufacturers so that they maintain needed capabilities in their integrated organizations and provide priority production as necessary.

A longer-term option would be to standardize parts and part numbers where possible, or even to standardize manufacturing technologies, part configurations, or component modules. Such standardization would mean that fewer parts would be needed in inventory and more parts could be available in each category. This strategy would require a very top level logistics approach to warfighting. While it would be possible to implement such an arrangement on a small scale, it is unlikely that many systems would readily sacrifice the functionality that comes with custom-designed components. However, if a standard complement of boards were designed and their availability was guaranteed and secure, it is possible that future system designs could make good use of them.

To truly mitigate the risk to DoD's mission, a decision tree for sustainment is needed. In that decisions are made, a decision-making method exists to a certain extent in the services and the DLA for legacy systems. New paths on this tree will need to be added in order to deal with innovation, future systems, and currently unpredicted supply-chain changes.

## THE DEFENSE INDUSTRIAL BASE

As the U.S. industrial base has changed, so has the U.S. defense industrial base. The defense product and systems integrators have consolidated, whereas the supply chain—which serves both defense and commercial products—has become increasingly distributed.

The commercial industrial base has become more innovative and faster moving; products are increasingly complex and depend on advanced electronics and other technology. At the same time, production has become more distributed. A more distributed industrial base means that technology no longer resides only in large centralized companies with major R&D laboratories.

A prime example of these changes is the role of small and medium enterprises (SMEs). Smaller companies, many with fewer than 100 employees, are now technology leaders in many emerging fields and represent some of the most active and innovative businesses in the United States. This trend is driven by the following:

- The ability of small companies to get sustainable funding through venture capital government funding;[22]
- The increase in outsourcing (but not necessarily offshoring) by large companies of research, development, innovation, and services; and
- The increase in the availability and capability of contract manufacturers.

SMEs typically provide capabilities that their larger customers do not have or cannot cost-effectively create, such as the following:

---

[22] Examples of appropriate government funding are SBIR (Small Business Innovation Research) and STTR (Small Business Technology Transfer) grants.

- Agility in responding to changes in technologies, markets, and trends;
- Efficiency due, in part, to less bureaucracy;
- Initiative and entrepreneurial behavior on the part of employees, resulting in higher levels of creativity and energy and a greater desire for success;
- Access to specialized proprietary technologies, process capabilities, and expertise;
- Shorter time to market because operations are small and focused;
- Lower labor costs and less-restrictive labor contracts;
- Spreading the costs of specialized capabilities over larger production volumes by serving multiple customers; and
- Lower-cost, customer-focused, and customized services, including documentation, after sales support, spare parts, recycling, and disposal.[23]

While SMEs have many advantages, one concern of DoD is that SMEs tend to form, fail, and change hands more often than large businesses do, so the technologies fostered by SMEs may be more easily lost. The United States has traditionally welcomed foreign direct investment and provided foreign investors fair, equitable, and nondiscriminatory treatment, with few limited exceptions designed to protect national security.[24] However, this regulatory framework for mergers and acquisitions does not have the same restrictions on small businesses, so key defense technologies may be easily acquired by foreign firms. In addition, the purchase of a leading small company producing a particular new technology by a defense prime product integrator can dampen the entire market. Because of the alignment and consolidation among the defense prime firms, the resulting lack of competition can lead to the purchase or failure of several similar companies and can eliminate future innovation.

A more intrinsic concern is that many small companies focus on a single application for an innovative technology. DoD is interested in ensuring the survival of a relevant embryonic technology and is also interested in the technology's ability to supply a multiplicity of applications that may exist in the defense enterprise. To foster such development, new defense funding models are needed. For example, under current regulations, a small firm can receive millions of dollars of federal funds for research, equipment, and personnel development, and then be sold, packed up, and moved to foreign soil.

Because defense acquisition processes are designed for large, dedicated companies, DoD is generally seen as a difficult customer for small businesses or commercial businesses. DoD seeks to expand the defense industrial base to commercial sources of technology—large or small. Today, few of these firms are seeking defense contracts despite the defense need for their products. Defense shares technology needs with the medical community, commercial information technology, and law enforcement, among others.

The demographics of the PrCB industry in the United States have changed, and most of the larger entities have closed, leaving a field predominated by SMEs. According to industry sources, the number of PrCB manufacturing plants is about one-third of what it was a decade ago. A few dozen of these companies are qualified to produce PrCBs for DoD. Logically, even with a smaller number of sources, it is possible to ensure price competitiveness and adequacy of innovation. However, the discussion here is meant to show that merely having multiple sources is not enough and that other factors must be considered.

A key issue is that the PrCB is a subtier-of-a-subtier product and is acquired many levels below the direct purview of DoD purchasing managers. DoD, therefore, has had little visibility into who makes which intermediate product for a pedestrian element of a subsystem that a prime contractor then integrates into a purchased item. Over the past decade, when the commercial PrCB business went offshore along with many other elements of commercial electronics, DoD was confronted with a new situation. It is now faced with the task of tracking something that it had not previously tracked in any way.

---

[23] National Research Council. 1998. Surviving Supply Chain Integration: Strategies for Small Manufacturers. Washington, D.C.: National Academy Press, p. 13.

[24] The 1988 Exon-Florio amendment to Section 721 of the Defense Production Act of 1950 authorizes the president to prohibit "any merger, acquisition or takeover" of a U.S. company by a foreign entity if "there is credible evidence that the foreign entity exercising control might take action that threatens national security."

## Buying American

While the Department of Defense does not exclusively "buy American," the Buy American Act of 1933 continues to carry the force of law in the United States. The act was passed with the purpose of providing preferential treatment for domestic sources of manufactured goods and construction materials. The act has been a source of some debate because of "its complicated nature, the requirement for certification of compliance by defense contractors, and its continued existence in an era of acquisition reform."[25]

Note that the Buy American Act applies only to end products and not to their components and subcomponents. "End product" can mean an unmanufactured end product that has been mined or produced in the United States, or an end product manufactured in the United States if the cost of its components mined, produced, and/or manufactured in the United States exceeds 50 percent of the cost of all of its components.

Policy makers recognize these challenges and bypass them by imposing domestic content ratios. The Buy American Act specifies that, unless exempt, government agencies must buy products for which at least 50 percent of the costs were incurred in the United States. Likewise, car companies selling more than 100,000 vehicles have minimum domestic content ratios. These policies attempt to define what an American product is, but their effectiveness is doubtful. They loosely address the locality of manufacturing but are not concerned with ownership—that is, where profits go—nor with office locations, in particular that of headquarters, where decisions are made.

As of a result of this legislation, companies can manufacture parts overseas and then assemble them in the United States into U.S.-made products. To meet the targets in the legislation, cost reports may be represented in a variety of ways. For example, the calculation may include full overhead in the U.S. plant but only labor costs for the overseas production. Congress has considered ways to strengthen these requirements, including an increase in the domestic content ratio in the Buy American Act to 75 percent. However, doing so will not resolve the definition of an American product or an American company.

The approach that Congress has taken to the question protects the manufacturing segment to an extent. The integration of components, subsystems, and systems has become an important function of the manufacturing industry. But according to that definition, Honda Motor Corporation is classified as an American company. And according to that definition, Iranian companies that comply with the standards can also be labeled as American. This raises security issues: the Buy American Act is particularly concerned with defense expenses. Can an Iranian company be a trusted supplier to DoD? Can a company owned and operated by antiwar extremists be a trusted supplier to DoD?

The popular view that DoD should buy American is not held by all. If foreign content is low and likely will remain so, limiting defense expenses to American products can be seen as shortsighted. If the highest-quality or most cost-effective or most innovative component is not made in the United States, it would be negligent of the defense acquisition corps not to purchase it for the warfighter.

The Buy American Act is only one of the dozens of rules that affect the U.S. military industrial base. These rules have all been implemented with national security, fair and open competition, and the U.S. economy in mind.[26] The Buy American Act provides very good examples of the intended and unintended consequences of such legislation. Acquisition reform has had added an additional layer of complexity, but it is difficult to tell if it has resulted in true improvement in an era of constantly shifting strategic challenges.[27]

The Buy American Act does not explicitly address the issues of either innovation or trust. The writers of the act may have intended this, but there is nothing in the procedures that address these issues; therefore they remain unaddressed in many cases.

---

[25] J.S. Smyth. 1999. The impact of the Buy American Act on program managers. Acquisition Review Quarterly Summer: 263-272.

[26] J. Spencer, editor. 2005. The Military Industrial Base in an Age of Globalization. Washington, D.C.: The Heritage Foundation, p. 11.

[27] C.A. Murdock and M.A. Flournoy. 2005. Beyond Goldwater-Nichols: U.S. Government and Defense Reform for a New Strategic Era. Phase 2 Report. Washington, D.C.: Center for Strategic and International Studies, p. 91.

## Global Companies and Their Complexities

The question of American products and American companies is a complex one. Globalization is abolishing the meaning of these terms to the point that any definition has become an approximation. While some people think of companies such as Intel Corporation or General Motors as U.S. companies, this may not be the case, because they are employing more and more people overseas and their customers are increasingly overseas. Most companies view it as their responsibility to their stockholders to source the most attractive resources around the world. It is difficult to know how a company's allegiance to the home country of its headquarters influences its decisions.

Looking at ownership can be a better way to determine what an American company is: it is in the interest of the United States as a whole to buy from companies that employ American workers and that have American owners, because profits will then stay in America. But for publicly owned companies, stock will not necessarily stay in American hands even if a company was originally wholly American-owned. Nokia Group represents another side of the same argument: while it is largely recognized as a foreign company, the majority of its stock is held by Americans.

Nokia is considered foreign because its headquarters are outside the United States, but all of the products that it sells on the American market are manufactured on American soil—whereas the American company Motorola, Inc., manufactures most of its telephones abroad. Whether Motorola is more American than another company merely because its headquarters are in the United States is a difficult question.

Many traditional U.S. companies now receive a majority of their revenue from foreign sources, and a majority of their employees, operations, and/or resources are overseas. Many of these companies never repatriate their foreign income, meaning that they yield little ultimate benefit to the U.S. economy. This perspective can lead to some interesting conclusions; a company such as Toyota may be incorporated in Japan at the parent level and may be culturally Japanese, but it may be as much of a U.S. company as other multinationals are. The multinational business community increasingly views itself in this light, as it is primarily in the business of increasing shareholder value.

> The question is what is a U.S. company? If we look into the law, a U.S. company operating within the boundaries of the United States, even if it is Honda Motor Corporation, is a U.S. company. Foreign corporations operating within the United States are defined as U.S. corporations because they operate within our soil.[28]

The most important consideration in this discussion is the lack of understanding of the relationship of these issues to the long-term economic stability of the United States. This view of the business world relies on a level playing field. A level playing field, while sounding desirable, really implies a reliance on markets and competence and a lack of reliance on statecraft and leverage that can leave national and economic security at risk.

## Policy Implications for PrCBs

It has become increasingly difficult to buy PrCBs under the provisions of the Buy American Act. First, the military dependence on COTS electronic systems for both new and old weapons and equipment is increasing. At the same time, the ability of U.S. companies to supply those technologies is lagging. Hong Kong, Japan, Taiwan, and the United States are investing millions of dollars in manufacturing capacity in China.

The existing printed circuit board manufacturers in the United States are small to medium-size enterprises. Along with this trend in company size, the military has seen increasing cases of single-source and no-bid conditions for some of its most critical leading-edge printed circuit board requirements.[29]

---

[28] Congressional Record. 2004. Statement by Congresswoman Marcy Kaptur, R-Ohio, in consideration of H.R. 1561, United States Patent and Trademark Fee Modernization Act of 2003. House of Representatives, March 3, p. H792.

[29] Presentations to the committee by representatives from Naval Surface Warfare Center, Crane Division (Indiana) and Warner Robins Air Logistics Center (Georgia). December 2004.

Former Deputy Under Secretary of Defense for Industrial Policy Suzanne Patrick has argued that a Buy American approach designed to strengthen the U.S. industrial base would adversely affect the defense supply strategy. Current studies carried out by the Department of Defense find no national emergency in acquisition policy nor the defense industrial base.[30] The prevailing policy for defense acquisition is currently to buy globally the best value. According to Ms. Patrick:

> From the extensive semiconductor study we have underway it appears that for most of the military requirements, we currently have a sufficiently large and robust semiconductor industry here in the United States—domestic capacities significantly exceed the demand for defense-specific semiconductors. The fact that our semiconductor companies are responding to global demand elsewhere by building additional facilities to address that demand only will strengthen the commercial viability, innovativeness, and pricing capability of those companies, to our advantage.[31]

While Ms. Patrick's comments are directed at the semiconductor industry, the policy applies to the entire defense industrial base. Again, a number of countervailing views are readily available. Most predominantly, the President's Council of Advisors on Science and Technology found that, "in the face of global competition, information technology manufacturing has declined significantly since the 1970s, with an acceleration of the decline over the past five years. While the U.S. has largely remained dominant in leading edge design work, U.S. industry experts are increasingly anxious over losing this advantage."[32] Again, this applies to all aspects of information technology—PrCBs, semiconductor chips, and supporting software.

DoD remains concerned about diminishing manufacturing sources and material shortages, defined by DoD as the loss or impending loss of the last capable manufacturer or supplier of raw material, production parts, or repair parts. Various strategies have been identified for addressing particular diminished and diminishing sources, and some are currently being implemented.

## Foreign Sources, Foreign Sales

An outdated paradigm is that the U.S. military is the primary and sometimes the only customer for many of the products made using the most advanced military technologies. In the case of the defense-unique equipment, the military has come to expect the companies that produce them for the United States to expand their sales to U.S. allies, thereby lowering the cost through economies of scale. A decade ago, defense suppliers were encouraged to work with foreign sources. By doing so, they brought better technology to DoD while saving the government money.

This strategy had the added benefit of allowing the government to create an international dependence on U.S. industries for global security. This is a very complex shift from the previous model, by which the U.S. government would defend a country or region directly with U.S. troops and U.S.-owned equipment rather than allowing the country or region to defend itself by purchasing U.S.-manufactured defense goods. However, shrinking global military budgets have reduced the demand for military equipment, creating a buyer's market. At the same time, increased international competition can allow buyers to negotiate very favorable deals with manufacturers.

This strategy is complicated by a number of factors. "Offset agreements" are side deals forged between the defense manufacturers and the countries purchasing their products. These agreements require a supplier to direct some benefits, usually component production or technology development, back to the purchaser as a condition of the sale.

Offset agreements can be a result of either defense or nondefense sales, and they can be direct or indirect. Through direct offsets, the purchasing country receives contracts for production or technology development directly related to the sale. Indirect offsets may involve unrelated investment in the buying country or the transfer of technology unrelated to the weapons being sold. Both types of offsets send

---

[30] Office of the Deputy Under Secretary of Defense (Industrial Policy). 2004-2005. Defense Industrial Base Capabilities Studies. Available at http://www.acq.osd.mil/ip. Accessed September 2005.

[31] Manufacturing News. 2005. What is the real health of the defense industrial base? An interview with Suzanne Patrick, Deputy Under Secretary of Defense for Industrial Policy. 12(4):1-9.

[32] President's Council of Advisors on Science and Technology. 2004. Sustaining the National Innovation Ecosystems: Report on Information Technology and Manufacturing Competitiveness. Available at http://www.sia-online.org/downloads/PCAST_Report_January_2004.pdf. Accessed October 2005.

work overseas, but direct offsets also raise serious security concerns, as they assist the development of foreign arms industries.[33]

It is true in many cases that companies are offshoring to the allies of the United States. The complexities of global conflict today, however, have led to ambiguous support from some of our traditional allies for current war efforts. This troubles some policy makers; some believe that even the closest allies of this nation may not adequately control the flow of technology to non-allies. There is currently no comprehensive monitoring of technology transfer to China via our allies, or through deemed exports.

## Trusted Sources

A number of concerns regarding "trustedness" have arisen in this new paradigm in the context of the criticality of security. The trusted-source issue has emerged in the past two decades as DoD has come to realize that it could not rely on defense-specific sources for defense-unique components. Because such sources were becoming increasingly costly and falling behind their commercial counterparts, DoD turned from highly regulated captive vendors to sourcing components from commercial industry. The prospect of some adversary's compromising the component increased, raising the level of concerns regarding trust and security.

For example, DoD must be concerned about potential for tampering, especially for components that are traceable to critical defense applications. Tampering is a broad term; it may range from intentionally poor manufacturing practices intended to cause early, random, or unpredictable failures, to sabotage that may include Trojan horses, trapdoors, triggered failures, tracking devices, or other hidden failure modes. The current ability to test, upon delivery, very complex components for their reliability to perform as required—with no additional hidden features—is difficult to gauge.

In a large system such as DoD, the responsibility for such testing may rest with the manufacturer, the system integrator, or the government. Unfortunately in many cases, such testing is not considered by DoD policy at all. This possibility introduces special concerns to the low-volume, high-mix production environment and highlights the need for explicit built-in measures for assurance and trustedness. Currently, a number of electronic components are produced in classified facilities, but the percentage of all electronics produced this way is very small. Production is made classified if a particular chip, PrCB, or associated section of software code contains explicit classified information. Most electronics contain some information related to their final use, if only the basic information of how many are being ordered and where they are being shipped. It may be prudent to keep this information within a trusted circle of suppliers even if it is not classified.

The controversy in this discussion lies in the numbers. Electronics may be used in battle, in direct analysis or support of battle situations, or in a variety of logistics, personnel, and other indirect support situations. The argument can be made that some or even all of these components are critical to battlefield success; the counterargument is that treating every light switch or fastener as critical, vulnerable, and under threat will add unreasonable costs to an already costly system. While it is possible to produce some, or even all, of these components as classified components to achieve assurance, this is untenable. The added and onerous production, distribution, operation, and logistics constraints and costs are so high that no system can or would want to use the components in anything but the most highly classified systems applications. Therefore, both the vulnerability and the criticality of the component must be assessed before a decision can be made in this regard.

A simple risk assessment can be made step-wise as follows.[34] The numbers in parentheses below refer to areas on the risk model in Figure 3-2.

- Identify components that are critical to a mission (1);
- Identify vulnerabilities in these components that exist regardless of threat exposure (2);
- Identify threats that exist regardless of component exposure to them (3);
- Separate critical components with known vulnerabilities, but without known threat exposure (4);

---

[33] W.D. Hartung and F. Berrigan. 1996. Welfare for Weapons Dealers: The Hidden Costs of the Arms Trade. New York: World Policy Institute.
[34] After the model proposed by the DoD Critical Infrastructure Protection Program.

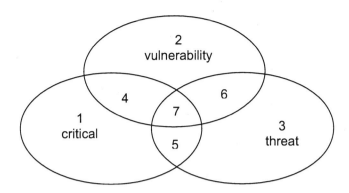

FIGURE 3-2  A simple risk model.

- Separate critical components with no known vulnerabilities, but with specific threat exposure (5);
- Separate noncritical components with known threats and known vulnerabilities (6);
- Protect critical assets with known, specific threats and known vulnerabilities (7); and
- Ensure regular, frequent, and repeated examination of the steps above.

With respect to the first item—identifying components that are critical to a mission—assessing the critical nature of a component can be done in a number of ways.  Consider that the definition of the word "critical" implies that the component is needed to avert a crisis.  In today's warfare situations, it is difficult to predict that one particular component could win or lose a war, but certainly any conflict is made up of a number of potentially critical events.  Therefore, the criticality of a component must be determined by the person or organization doing the assessment.

Regarding the second item, assessing vulnerability involves its own complexity.  Sources of the tools of warfare can be vulnerable to a number of factors—for example, a manufacturing plant that is a single source for a component could be shut down because of a natural disaster, a terrorist attack, or a labor strike.  Note that these events may not affect the plant directly, but could affect power, water, or subcomponent suppliers with the same effect.  A source might also be cut off because of a parent company's business decision not to supply the military, or because a company or plant ceases operation, or simply because of a failure to meet DoD quality standards.  An additional potential vulnerability is that exporting military designs may allow others to gain a technology advantage or to match the perceived U.S. advantage.  All of these factors, however, can be assessed.

In addition, a vulnerability assessment needs to consider the potential for tampering.  This can happen at the manufacturing plant at the component or subcomponent level, at any place in the supply chain.

Regarding the third item, assessment of threats: If there is no threat, then the vulnerability may never be exposed.  Simply being dependent on a foreign source, as an example, does not imply vulnerability.  Many foreign sources are very reliable and are long-term partners in global security.  However, dependence is worth tracking in order to understand vulnerability.  Critical vulnerability in this sense can describe a potential to lose a sole source of a material or component or to have that component compromised.

Once components are adequately labeled, those parts that do not fall into the danger zone where criticality, vulnerability, and threat coincide, can be procured anywhere reliably.  In the end, components that are critical, vulnerable, and under threat will be a small fraction of the total.  These will cost more per part owing to the security value, rapidly variable quantities required, and consequences when the part is not available.  But such risk is rarely included in the procurement decision, except in classified cases.

This approach implies that dependence on any source or set of sources is not in and of itself something to inherently avoid.  While a sole source implies vulnerability, reliance on a variety of foreign sources may be very robust.  In addition, reliance on foreign sources can be part of coalition warfare and other forms of well-structured statecraft, and these have had substantial benefits. Shared dependence through shared economic and global security goals can be shown to improve U.S. national security in many ways. According to a recent DoD report:

> Part of a DIBCS [Defense Industrial Base Capabilities Study] assessment is to evaluate how domestic industrial capabilities compare with foreign capabilities. This is necessary because, in order to provide the best capability to the warfighter, the Department wants to promote interoperability with its allies and take full advantage of the benefits offered by access to the most innovative, efficient, and competitive suppliers—worldwide. It also wants to promote consistency and fairness in dealing with its allies and trading partners while assuring that the U.S. defense industrial base is sufficient to meet its most critical defense needs. Consequently, the Department is willing to use non-U.S. suppliers—consistent with national security requirements—when such use offers comparative advantages in performance, cost, schedule, or coalition warfighting. For this reason, the Department and many friendly governments have established reciprocal procurement agreements that are the basis for waiving their respective "buy national" laws and put each other's industries on par as potential suppliers.

> U.S. sources for those technologies and industrial capabilities supporting warfighting capabilities for which it has established leadership goals to *be ahead* or *be way ahead* of potential adversaries could reduce certain risks associated with using non-U.S. suppliers. However, the Department must be, and is, prepared to use non-U.S. suppliers to support critical warfighting goals when necessary and appropriate, and when the supplier and the nation in which it resides have demonstrated reliability in:

> - Responding to DoD technology and product development requirements.
> - Meeting DoD delivery requirements during peacetime and/or periods of conflict or international tension.
> - Precluding unauthorized transfer of technical information, technologies, or products within the nation or to third parties.[35]

What is not possible for any component, therefore, is to rely only on COTS for both routine and specialized capabilities.  DoD has few internal capabilities left for developing new technologies, and it rarely invests in internal specialized needs.  These capabilities in many areas are not adequate for assessing the state of commercial technology or for assessing the technology needs of the warfighter. The intelligence community has pushed for higher internal capabilities and trusted foundries for microchips.  At the same time, DoD has shied away from the most cutting-edge technology in order to avoid risk.  While many needs may be met with commercial sources, it is unknown where the lines are best drawn, and there is no existing policy to create this knowledge.

Because of the commitment of DoD to performance-based contracting and the sustained downsizing of the DoD acquisition workforce, the ability to understand the manufacturing base, and the potential for technology development and application, many of these concerns are never adequately addressed, and certainly never on a systematic basis.  Understanding what is needed in a source is difficult in today's global manufacturing environment.  The definitions of a trusted source, a qualified source, and a reliable source are evolving.  Currently, DoD does not appear to have an adequate methodology to understand how any of these definitions should be applied to individual PrCB components.

## Conflicting Requirements

In the course of the debate over the role of the defense industrial base, two divergent opinions have emerged.  One view is that the vast majority of dollars collected through taxes should be spent to benefit U.S. taxpayers.  This may be through procurements or other spending at U.S.-owned companies and in U.S.-based facilities, to pay salaries in the United States, and to reinvest in infrastructure to support this base.  Restrictions on trade such as export controls, military critical technologies, intellectual property controls, and directed tariffs are seen as ensuring these benefits to the taxpayer.

---

[35] Foreign Sources of Supply: Assessment of the United States Defense Industrial Base.  2004. Available at http://www.acq.osd.mil/ip/docs/812_report.pdf.  Accessed October 2005.

The opposing view is that the United States is part of a global economy and that spending globally will always come back to benefit the United States. This view holds that although there may be an imbalance in certain commodities, industries, or locations in the short term, the overall result of trade is a benefit to the U.S. taxpayer. This benefit may be in the form of low-cost imported consumer goods; improved technology for U.S. energy, communication, and transportation systems; and exciting global opportunities for U.S. workers.

Defense acquisition seems to be in a transition period, from the past to the future, but different parts of the defense supply chain have very different views of the future. For example, the government has gone only part of the way toward adopting commercial acquisition practices. Commercial companies have come up with very effective ways to manage outsourcing, but the government still relies in many ways on specifications (admittedly performance specifications rather than military specifications) to guarantee quality.

One major commercial practice that has emerged during the shift toward outsourcing is to minimize risk through supply-chain management. While many models and theories for supply-chain management exist, few fit the military model, and so adopting the way that cellular telephones or automobiles are managed can be problematic. DoD rarely purchases more than a few thousand items at a time, and many purchases of large items (vehicles and artillery) are only in the hundreds; replacement parts are often in the single digits. A better model than consumer electronics, perhaps, is that used for biomedical devices or for high-rise elevators. These industries have similar safety, volume, and complexity constraints, and an integrator must be able to know who makes every part of its system, down to the circuit card and beyond.

Good commercial practice is to follow and understand one's supply chains in order to minimize risk. When the government does not do this, systems may need to be redesigned more often than previously, because it is easier to replace a large subsystem than individual parts. However, because the government has delegated the responsibility for system integration to a handful of prime contractors, the best interest of the government is not always driving the decision, or even well articulated or understood. And because the assumptions underlying these decisions are not well understood, these conflicting priorities are almost ubiquitous.

A number of additional reasons have slowed progress in DoD toward efficient procurement practices that truly integrate commercial and military manufacturing. Culture and history are most often cited, and the evidence for these are clear. Defense acquisition uses a system very different from that of commercial companies—everything from accounting practices to acronyms and jargon tends to be unique to defense procurement. This means that supply-chain managers who are uncomfortable with change will look first to historical sources of defense materiel.

The political reality of defense procurement is also important. True integration very often has led to charges that defense purchases can subsidize commercial product development and that this may upset free and fair trade. In addition, a number of high-profile charges of conflict of interest have exacerbated the confusion and have resulted in a retreat to heightened separation.

An obvious conclusion is that DoD must redress its hands-off role and provide more systematic oversight of requirements and acquisition. The move to commercial practices can work, but only with a commitment to success including a clear definition of exceptions. Can the government move faster to reach a real transition? Yes, with a strategy that analyzes and understands dependence, vulnerability, and critical vulnerability.

An interesting additional observation is that many commercial supply-chain practices are moving toward those of DoD in the days of military specifications. The level of tracking that was required by DoD contractors in the era before performance-based contracting was very rigorous, adding substantial cost to many products. However, this tracking was very valuable and has been found to result in very high quality and consistency. In the commercial world, many factors are driving OEMs toward better production tracking; among them are more geographically distributed outsourcing, which means that in-person process oversight does not happen, as well as evolving environmental regulations and the expectation that the OEM is responsible for the environmental footprint of a product. But the biggest driver for improved production tracking is the availability of new technologies, sensors, and controls that can accomplish this level of tracking at a very low cost. This type of tracking is fast becoming something that OEMs will be able to do remotely and globally with little additional cost. However, it is important to remember that tracking does not mean that quality is understood or integrated.

## KEY FINDINGS AND CONCLUSIONS

The committee finds that without publishing or legislating a defense industrial policy, the United States has a de facto policy, one that is in many ways inconsistent, is not well understood, and has a number of unintended consequences. This confusing array of current requirements does not allow the government to obtain the best value for the warfighter.

It is clear that DoD cannot—and should not—buy the more than 8 million different items that it needs from trusted sources. Nor should the government own the facilities to make all of these parts. There is a clear benefit to purchasing many of these from the private sector. However, some parts clearly should be procured from trusted sources. This is complicated because DoD rarely requires more than a few thousand of any one part, meaning that the number of replacement parts needed is often in the single digits.

A top-level need exists for finding a robust way to categorize parts, including components that warrant special attention for U.S. national security. This assessment must consider that there is a spectrum of needs and that there is no one-size-fits-all solution. Because producers and DoD program managers may be reluctant to incur the costs of such identification, this strategy must also have a way for program managers to incur the costs of their part numbers being so identified.

In conjunction with such a strategy, procedures are needed to develop and provide ongoing support for production of such critical components, either through subsidies to commercial businesses, through the creation of joint ventures, or through the use of a government manufacturing facility. Before this decision can be made, the ability and willingness of the commercial sector to respond to these needs must be assessed, as well as the ability and willingness of government facilities to respond. This strategy should also include an effort to incentivize industry in areas that matter to business and that do not encourage inefficient processes.

Government-owned production, such as that located on a military reservation or at a national laboratory and operated by government or contractor personnel, could be an attractive and potentially cost-effective option. This would be most appropriate for legacy parts, such as PrCBs for the SLQ-32 Electronic Warfare System, for parts with classified elements, or for any part for which assessment results in critical vulnerability.

These options appear to be desirable compared to a trusted production agreement with a commercial vendor. A government-owned facility has some well-documented downsides, including cost, the permanent nature of civil service employment, fewer incentives for improving productivity, and little opportunity for cost sharing with commercial production. An ideal government-industry partnership network would have to consider such key factors as cost, responsiveness, and contribution to innovation. A final consideration is the need for the government to maintain some capability to be a smart buyer.

Current program management and defense contracting practices do not appear to support the limitations imposed by a sole-source trusted supplier. To use a sole-source trusted supplier effectively, DoD procurement managers would need to apply judgment about matters on which they are currently not prepared, educated, or paid to do. However, absent a set of better analytical tools, a trusted supplier may be the only viable choice.

# 4

# Printed Circuit Technology Assessment

Electronic equipment is becoming pervasive and ubiquitous in all aspects of human endeavor, including military activities.[1] As electronics become increasingly integrated, the need to interconnect them will likewise continue to grow—and thus the demand for an increasing variety of printed circuit boards (PrCBs) entailing a variety of materials, form factors, and technologies.

PrCBs are sometimes considered as an older technology, the use of which is declining in current and future defense applications. This is simply not the case. The transformation of the armed forces under way in the Department of Defense (DoD) to become a more connected, electronic, and instrumented fighting force means that DoD now buys more PrCBs than ever before, for almost every component, subsystem, and system in use. In the past, many systems were free of electronics; in the near future, everything that a soldier uses, from clothing to food, may have integrated sensors. Of course, PrCBs are also key to electronics hardware used in communications, computer-controlled systems, electronic countermeasures, and fire control and avionics systems. All of these electronics need interconnection to operate, and therefore all will require a variety of printed circuit boards.

Because of this amazing range of applications, PrCBs are a complex and widely varying technology, ranging in size from millimeters to tens of centimeters. The thickness, layers, types, and numbers of interconnects, materials, and chemistry mean that the full range of capabilities and technology is difficult for any one facility to be able to produce. This complexity is very attractive for military performance in that it gives designers many options; it also can translate to headaches for any logistics manager with many different parts to track and maintain.

Because the military must maintain legacy systems, the range of complexity needed to fulfill past, present, and future needs grows exponentially. While the level of current technology on today's military systems lags a generation or two behind commercial technology, the current rate of change means that military acquisition will have to move much faster to stay within reach. The ability to manufacture very old technologies can be as difficult as it is to manufacture the very newest. As standards change, keeping older equipment operating presents many difficulties.

This entire picture is overlaid with the need for more secure, robust, and reliable PrCBs for military use. New applications will require longer life, reliability under more demanding environmental conditions, protection against tampering before their delivery to DoD, and increased security so that they can continue functioning even if attacked by state-funded hackers. The future needs for military PrCBs are difficult to gauge; the only incontrovertible fact is that more demands are coming.

---

[1] National Research Council. 2001. Embedded Everywhere: A Research Agenda for Networked Systems of Embedded Computers. Washington, D.C.: National Academy Press.

## WHITHER NEW TECHNOLOGY?

The military will need to expand the components that it purchases and stocks today to encompass a number of new technologies in PrCBs. The drivers for innovation in military systems, however, must be balanced with what is possible given new commercial manufacturing and materials innovations. For example, scenarios to standardize parts and limit configurations are possible, which means that more flexible programming options and standardized board designs are of some interest.

Military systems that depend on electronics hardware will increasingly require such emerging commercial interconnection technologies as liquid crystal polymers (LCPs), embedded microelectromechanical systems (MEMS), buried optics technologies (optoelectronics), and high-frequency substrates. The areas of impact of new PrCB laminate material development and application range from optics technology advances to microsensor technology development for chemical and biological detection and threat reduction. For improved reliability, security, and performance, improved sensors and process capabilities are needed in a number of areas.

Strong partnerships between government, industry, and academia are critical for innovation in this area. Such a collaboration of partners leverages a variety of otherwise unavailable experiences and capabilities and combines the resources to accelerate and increase the impact of the technology advancement. Technology changes will be necessary for the PrCB and interconnection technology to meet the dual mission of providing for legacy product requirements and equipping the warfighter with new hardware. Anticipating areas and directions of technology emergence in future critical technologies can enable the accelerated deployment of new technology as well as more-effective support of that technology for sustaining the warfighter.

## U.S. INDUSTRY RESEARCH AND DEVELOPMENT

Historically, the original equipment manufacturers (OEMs) funded much of the research and development (R&D) for the PrCB market. The OEMs are the companies that sell the final assemblies that incorporate PrCBs; historically they also manufactured the PrCBs. Many OEMs believed that the PrCB was of fundamental importance, and that by maintaining manufacturing as part of their core competency they were better able to drive the technological advancements needed for their specific equipment. In addition, many OEM companies funded R&D because they had sufficient overhead and personnel time to make these types of investments; they also had the best idea of what type of performance would be needed and the best understanding of design constraints for next-generation electronics components.

In 1980, 52 percent of the PrCBs manufactured in the United States were made by OEMs and their captive manufacturers. These OEMs traditionally spent approximately 10 percent of the sales value of these in-house-produced PrCBs in R&D efforts to improve manufacturing, quality, and yields. By 2001, in contrast, the committee estimates that the percentage of PrCB production by OEMs had dropped to 1 percent. And by 2004, only a few OEM facilities remained, with a capacity (primarily for dedicated military products) that was less than 0.1 percent of the total U.S. output. As a result of this shift, the traditional sources of R&D funding dropped by two orders of magnitude. In reality, the critical mass of R&D in this industry disappeared, reducing the investment in new technology to near zero.

Another large source of R&D funding and activity previously came from the supply base. As OEMs sought and turned to outside sources for PrCBs, a subtier supply industry emerged. Initially, the supplier industry funded a level of R&D similar to that of the OEMs as a percentage of sales, but this has changed for U.S. suppliers. Funding for technical activities from these manufacturers was once estimated to be 10 percent of all U.S.-generated supplier sales dollars in the 1990s. By 2001, both the level of sales and the percentage spent on R&D was decreasing. In 2005, it is estimated that less than 3 percent of sales is spent on technical activities to support PrCB manufacturing. The effect of this loss of R&D spending is expected to have long-term effects. In 1999, U.S.-based PrCB suppliers spent an estimated $50 million on technical activities and new-process and -product R&D. In 2005, this sector will spend less than $10 million for such R&D.[2]

---

[2]  These estimates were gleaned from discussions with industry experts at the committee workshop held December 13-14, 2004. See Appendix D for a list of attendees.

One model of innovation is the supplier contributions to R&D. Over time, these have resulted in many new products and processes. In general, the contributions by the OEMs have resulted in improved process efficiencies. The greater-than 75 percent reduction in revenue directed toward new-product and -process R&D could easily result in stagnation of innovation in this sector in the United States, with the outcome that U.S.-based PrCB manufacturers will fall farther behind on the global technology curve.

In some areas, military and commercial PrCB technology is currently going in different directions. For example, military PrCBs are combining radio-frequency (RF) and digital technology and are combining more functions in a single board. The design of military PrCBs is driven by spiral development, which can keep military features out of step with commercial boards; the latter are designed from the ground up when improvements are made.

## GLOBAL RESEARCH AND DEVELOPMENT

Today, most notably in Korea and Japan, non-U.S. captive OEM PrCB manufacturers continue to fund innovation. Companies including Samsung Electronics, LG Electronics, and NEC Corporation have been very active in R&D for PrCBs; this effort has resulted in advancement for the respective companies' specific product requirements, a better position for new-product innovation, and a significant strategic competitive advantage. Many U.S.-based companies once espoused this business strategy, but the majority of the PrCB industry, along with many U.S. manufacturers, now treat manufacturing as a static, commoditized process rather than as a product-development opportunity.

The availability of government funds that support R&D efforts at both small and large companies is small when compared with that available under the older business models. These funds, including grants, subsidies, and tax incentives, can be used to support development activities focused on next-generation PrCB manufacturing processes and electronic designs. The struggling industry, from the largest OEM down to the smallest PrCB shop, can no longer justify support of U.S.-focused R&D to its owners or stockholders. Other nations provide greater financial support for R&D to the OEMs, which in the past have shouldered the burden of product R&D.

Because of the relative abundance in government R&D support outside the United States, major electronics OEMs are building R&D facilities outside this country. This investment takes current and future technology development outside the United States, but it also affects current and future manufacturing employment. Among the OEMs building such facilities are Cisco Systems, Inc., and Intel Corporation, which both announced large investments in R&D centers in China in 2004.[3]

Partnership efforts—in such organizations as the National Institute of Standards and Technology (NIST), the Defense Advanced Research Projects Agency (DARPA), and the former National Electronics Manufacturing Initiative (now iNEMI)—can only begin to fill the U.S. R&D gap left by the OEMs. In some cases the technology partners involved in these U.S. industry- and government-sponsored consortium activities are anything but U.S.-centric.

DoD states that "electronic systems and subsystems represent about 40 percent of the defense acquisition budget and are the critical enabling technology that differentiates our weapon systems. As the Department of Defense downsizes, it has become increasingly important to keep access to affordable, advanced electronics technology by leveraging the high volume, leading edge, merchant manufacturing infrastructure . . . ."[4]

This high-volume, leading-edge, merchant manufacturing infrastructure, however, is no longer vibrant in the United States. What does this mean for the future of the nation's electronic systems? Will

---

[3] From *Electronics News,* September 23, 2004: "Cisco said it would invest $32 million in its R&D center over the next five years and expects to hire about 100 employees over the next 18 months for the site. The company believes this R&D center will enhance its ability to tailor products and be more responsive to changing service provider customer demands in China, Asia and around the world." They are definitely not focusing on U.S. requirements in this Chinese R&D facility. And from *Sci-Tech,* a Chinese electronics newsletter of April 29, 2004, "Intel Corp., the world's largest computer chip maker, will set up a research and development center in Shanghai, with an investment of US$39 million, according to a document signed Thursday. The R&D center will focus on the development of chip-related products and on customer services, said Siew Hai Wong, vice president of Intel, adding that the new research facility would provide 450 jobs at the first phase."

[4] Commerce Business Daily. 1996. Electronic systems manufacturing and design support for mixed-technology integration. Issue No. PSA-1552, ARPA Broad Agency Announcement 96-16.

the United States lose the competitive edge that has so far been critical to differentiating its warfighters? The government has sought, through targeted solicitations, to alleviate barriers by increasing the flexibility of the merchant electronic services manufacturing (ESM) infrastructure and streamlining the process by which new products are designed and transferred to manufacturing. This implies a mission that would lead to a vibrant program with hundreds of active projects in the area of electronics systems, available for participation by U.S. research and manufacturing firms. Instead, programs are on the decline. For the remainder of 2005 and into early 2006, only 13 projects are active, and all but one program is closed to solicitations. There is no indication that this trend will be reversed.

## TECHNOLOGY CONCERNS

Some of the overarching technology concerns for the military focus more on processing advances than on technology performance. For example, cost-effective, low-volume manufacturing could have a larger impact on overall military effectiveness than would a faster interconnect on a single PrCB. Improved configurability could also make a big difference in overall effectiveness. Related technologies might include more agile manufacturing, allowing many configurations to be made on single product lines, and more robust manufacturing to ensure that reliability comes with increasing agility.

Another pressing overarching concern is for more environmentally benign manufacturing, including the reduction or elimination of chemical waste. It is also anticipated that DoD will need to make both PrCB processing and disposal lead-free. This change may come about in response to regulations, public pressure, or the eventual implementation of lead-free electronics as a global standard.

However, the most pressing concern is that the United States may not have access to better technology than its adversaries have. This is the concern that drives the military and industrial suppliers to develop better designs and therefore to hire better designers. Many argue that this knowledge is the source of the real intellectual property advantage and should be where the investment is concentrated. However, current practices, including the drive toward commercial and military industrial base integration, mean that control of this knowledge is very difficult. Slowing the spread of the knowledge may be possible, but with an economic penalty. The only way to maintain an advantage is to stay ahead—which means keeping a high investment in research and not merely funding the development or transition of commercial technologies.

The foregoing discussion implies that there are two overall technology approaches. First, it may be possible to determine what technology is most strategic or is likely to provide an asymmetric advantage. While this is certainly possible, it is not a guarantee against technology surprises. A second option is to "replenish the pool" internally. This can be achieved only with investments in basic technology, in product R&D, and in process R&D. With this option, the goal is to determine "where to run faster." This approach encourages spending money on the most difficult and important challenges. This is where the United States excels. This approach approximates a defined and measurable technology policy in that it sets goals and provides incentives, and disincentives, that will lead technology down a desired path.

## POTENTIAL APPROACHES TO SUPPORT TECHNOLOGY INNOVATION

According to the vision of the National Innovation Initiative, "Innovation fosters the new ideas, technologies, and processes that lead to better jobs, higher wages and a higher standard of living."[5] It is generally accepted that the United States has a record of sustained innovation over decades, across industries, and through economic cycles. The situation for PrCBs reflects this trend in every respect. The relevance of this sustained innovation for DoD has increased as DoD has grown to depend more and more on commercial technology, manufacturing, and R&D. If that commercial technology starts to lag behind, DoD must take action or suffer the consequences. In addition, other concerns are arising; some of these are addressed below.

---

[5] Council on Competitiveness. 2005. National Innovation Initiative. Available at http://www.compete.org/nii/. Accessed October 2005.

## Technology Approaches

The vast majority of PrCBs are manufactured by plating a layer of copper over one or both sides of a substrate, and then removing the unwanted copper and leaving only the copper traces as the interconnection pattern. Some PrCBs have a trace layer inside the PrCB (multilayer). After the circuit board has been manufactured, components are attached to the traces by soldering. Three common methods are used for the production of printed circuit boards:

- *Photoengraving* is the use of a photomask and chemical etching to remove the copper foil from the substrate. The photomask is usually prepared with a photoplotter from data produced by a technician using computer-aided PrCB design software. Laser-printed transparencies are sometimes employed for low-resolution photoplots.[6]
- *PrCB milling* is the use of a two- or three-axis mechanical milling system to mill away the copper foil from the substrate. A PrCB milling machine operates similar to a plotter, receiving commands from the host software that control the position of the milling head in the x, y, and (if relevant) z axes.
- *PrCB printing* is the use of conductive ink or epoxy to form traces directly on substrate material.

While these methods and materials are currently the most practical, new concepts could improve the design, configurability, accuracy, complexity, reliability, and cost of PrCBs.

Some new technology trends include new materials, including liquid crystal polymer substrates, high-frequency laminates, and new microwire materials. Some new processes will certainly also be adopted for military use, although the time line will most likely be one or two generations behind commercial applications. These processes may include optical backplanes, three-dimensional printing, wireless sensors, embedded passives, metal potting of components, and high-frequency technology. DoD needs to plan for ensuring access to these technologies and to guard against tampering.

## Regulatory Approaches

It is well understood that better life-cycle management of electronics systems can be achieved through the elimination of lead in the solder, coatings, and components on PrCBs. The cost of such a change, however, means that the transition must be carefully coordinated. Standardized processes throughout the supply chain and the OEMs have also precluded any single company from moving toward a different, no-lead standard.

However, recent European regulations to restrict hazardous substances are now leading a global trend toward the elimination of lead in many processes. While the U.S. government does not currently regulate this standard, some states are doing so, and it is anticipated that the global nature of the industry will cause such a change in the near future. Military products are currently exempt from the restriction of hazardous substances (RoHS) directive (as discussed in Chapter 2), but it is understood that many suppliers will not want to run separate production lines and that most will therefore migrate to lead-free production.

No-lead reliability is a concern for military components for a number of reasons. Of particular concern is the change to higher reflow temperatures than that which lead-free solders generally require. Lead-free products have been on the market for several years using solders containing bismuth or zinc, with melting points close to 200°C. It is now clear that the most likely lead-free solders for wide-scale use are those melting at about 220°C. This change on both components and assembly will require a requalification of the entire production process as well as requalification of a subassembly and eventually of entire systems affected.

A major concern is the lack of data on the long-term reliability of products using lead-free solders. Many commercial companies are reluctant to guarantee products made with the new standards for more than a few years. The military often requires components with much longer life, and few data currently exist to support such long-term reliability.

---

[6] Additional information is available at http://www.fullnet.com/u/tomg/gooteepc.htm. Accessed October 2005.

The use of lead-free processes will affect several aspects of PrCBs, including the finish on component lead frames, the joints that connect components to the board, and the plating on the board itself. The effect of the high-temperature excursion during lead-free soldering on the long-term performance of active and passive components has been investigated, but more work on the long-term performance is needed. Concerns range from the immediate effect of the high-temperature lead-free assembly process on temperature-sensitive components, to the long-term life of all components after passing through a high-temperature assembly process, to the effect that a specific mix of PrCB components has on overall reliability. The susceptibility in components with pure-tin finishes to the formation of tin whiskers is also a growing concern.

While the higher-temperature technologies and no-lead components are in everyday use in many applications, as the implementation of lead-free solders increases across commercial industry, the labeling, shipping, and managing of components are becoming a concern. Because some components could exist in two variants—one to comply with lead-free applications and the other for previously specified leaded applications—confusion in the supply chain is inevitable. The crux of the problem becomes one of extended, global supply chains that will make it more difficult to ensure compatibility for critical systems.

Finally, a number of other existing environmental regulations and international standards can introduce additional complications. These include worker safety and the disposal of the many chemicals used in the PrCB manufacturing process.

## Organizational Approaches

A number of organizational approaches have been tried for improving the research and innovation capacity for similar technologies. An industry consortium called the Interconnect Technology Research Institute (ITRI) was formed in 1994 specifically to cooperate on R&D for PrCBs. A group of manufacturers and suppliers came together to try to fill the growing gap in technological competitiveness of U.S. PrCB production. This organization was chartered to facilitate North American technology advancements in the area of PrCB manufacturing. Although the group operated for 6 years, declining PrCB manufacturing participation, especially by OEMs, caused ITRI to close in 2001.

In 1998, the PrCB industry approached the U.S. Congress for the initiation of a center to address lack of U.S. research and development in board technology. The PWB (Printed Wiring Board) Manufacturing Technology Center (PMTEC) was then formed to address the development of state-of-the-art PrCB technology. The center was operated by an arm of the Illinois Institute of Technology Research Institute (IITRI) based in Huntsville, Alabama; it is now called the Alion Science and Technology Center. PMTEC partnered with a commercial manufacturing company to ensure an affordable, responsive, and reliable U.S. PrCB manufacturing capability to meet current and future DoD requirements. The effort was executed through an integrated program of research, education, and technology transfer. PMTEC focused on propelling the development of bareboard technology; it supported current and future advances in packaging and PrCB assembly technology. The effort also addressed unique and critical military PrCB needs, such as the ability to withstand harsh environments, long-term availability and reliability, rapid insertions, and integration of new technology. Although PMTEC successfully teamed with an industrial producer, this partnership was not sustainable under commercial pressures. Eighteen projects were completed through the center while it received DoD earmarked funding between 1998 and 2003.[7]

By comparison, the semiconductor industry has been more successful in sustaining both industry and industry-government consortia. The best known of these is the SEmiconductor MAnufacturing TECHnology, or SEMATECH, an experiment in industry-government cooperation conceived to strengthen the U.S. semiconductor industry. The consortium was formed in 1987 when 14 U.S.-based semiconductor manufacturers and the U.S. government came together to solve common manufacturing problems, improve the industry infrastructure, and work with domestic equipment suppliers to improve their capabilities.

By 1994, the U.S. semiconductor industry had regained strength and market share, and the SEMATECH board of directors voted to seek an end to matching federal funding after 1996. SEMATECH

---

[7] Printed Wiring Board Manufacturing Technology Center. Summary available at www.armymantech.com/success/pmtec.pdf. Accessed October 2005.

continued to serve its membership and the semiconductor industry at large through advanced technology development in program areas such as lithography, front-end processes, and interconnect, and through its interactions with an increasingly global supplier base on manufacturing challenges. However, it did little to meet the DoD's need for affordable, low-volume production. In 1999, SEMATECH renamed itself International SEMATECH. It is now a unified global consortium, with members from Asia, Europe, and the United States, dedicated to cooperative work on semiconductor manufacturing technology. It has no U.S. government nor DoD focus or support, and it no longer addresses defense needs or reliability.

A slightly different approach was taken to form the Defense Microelectronics Activity (DMEA). This organization began in 1981 as a small unit in the Engineering Division of the Sacramento Air Logistics Center at McClellan Air Force Base near Silicon Valley. Initially called the Advanced Microelectronics Section, its start coincided with the incipient use and growing necessity of microelectronics in weapons systems. The unit was created to assist the U.S. Air Force, but eventually it came to serve all of the Department of Defense. In 1997, the unit moved from the administrative structure of the Air Force to the Office of the Secretary of Defense.

The DMEA has the mission of addressing the growing problem of microelectronics obsolescence. Several parallel approaches to this problem include accessing and storing the design drawing and processes to make legacy parts. The DMEA can also procure parts and supply them to military units, and in some cases can manufacture microelectronic parts on demand. The DMEA also has the expertise in developing new microelectronic technologies, including the ability to design, prototype, and test components and systems. This capability to design, prototype, and test does not currently extend to PrCBs. The DMEA is not a consortium and does not partner with industry; it is fully funded by DoD.

A third example is the International Electronics Manufacturing Initiative (iNEMI), an industry-led consortium, with no DoD funding, whose mission is to assure leadership of the global electronics manufacturing supply chain. With a membership that includes approximately 70 electronics manufacturers, suppliers, associations, government agencies, and universities, iNEMI provides an environment for partners and competitors to anticipate future technology and business needs collectively and to develop collaborative courses of action to meet those needs effectively.

Given this wide variety of organizational attempts at improving research and innovation capacity, the PrCB industry should be able to identify some key program approaches that will lead to a successful research effort. For example, these attempts demonstrate that the ability to change and remain relevant with changing external pressures can help to maintain a steady source of support. They also demonstrate very well that commercial interests alone cannot fulfill DoD needs.

One major difference between PrCB fabrication and integrated circuit fabrication is the capital equipment cost. The cost of equipment to recapitalize board manufacturing is estimated to be less than $10 million per year, which is orders of magnitude below the $1 billion estimates for building a state-of-the-art microelectronics foundry.[8] Therefore, it may be cost-effective for DoD and its U.S. supplier base to maintain fabrication technology competence at levels needed both to support legacy production and to carry out R&D for future needs.

One existing effort that is attempting to fill this need is the Emerging Critical Interconnect Technology (E/CIT) effort, established currently with a small PrCB manufacturing capability sited at a military base. The E/CIT program activity is coordinated through the Printed Circuit Technology Branch of the Naval Surface Warfare Center, Crane Division, located in Crane, Indiana. The facilities are available to joint development projects that support technology advancements needed by the domestic PrCB military and commercial industry. The site limitations could be used to restrict use by non-U.S. personnel, which could effectively limit use to U.S. companies and their U.S. employees—presumably precluding any inevitable addition of the word "international" to the name of the center.

Such a facility has the potential to provide utility to both the military and the commercial industry. As has become clear, it is unlikely that pointing aid solely at either commercial or military entities will provide a sustainable path to DoD. If such a joint facility were supported, it could potentially maintain and perhaps even outpace global technology competence. The E/CIT claims that it can provide a low-cost access for U.S. companies to R&D in this critical area.

---

[8] These estimates were gleaned from discussions with industry experts at the committee workshop held December 13-14, 2004. See Appendix D for a list of attendees.

## KEY FINDINGS AND CONCLUSIONS

History indicates that innovation is important to meeting both legacy and future DoD needs in interconnection technologies. Current R&D funding, however, from either industry or government sources in the United States is not now adequate to ensure U.S. access to leading technologies. Given current trends, it is conceivable that our adversaries will be able to access some of these advanced technologies more easily than the United States government and suppliers.

There are no simple solutions available to remedy this situation. Any approach that is considered should take technology, regulations, and organizational considerations into account.

# 5

# A Systems Approach

Printed circuit boards (PrCBs) are a key technology for the current and continued success of nearly every intelligent system in use today—consumer electronics and telecommunications come immediately to mind, and transportation, energy, and infrastructure follow quickly. The dependence on PrCBs of systems for national and homeland security is also very clear. The transformation of today's fighting force to a more connected, electronic, and instrumented fighting force means that the Department of Defense (DoD) now buys more PrCBs than ever before, for almost every system in use.

The technology to manufacture these components and systems is becoming increasingly complex. New technologies for substrates, interconnection technologies, packaging, and design are enabling a new generation of devices and electronics. Yet for many commercial applications, costs are going down as complexity goes up. The cost of the newest technologies is now driven not by military technology pull but by commercial push and, inevitably, by the globalization of the industry surrounding this product.

After they are fabricated, printed circuit boards are populated with electronic components and integrated circuits, and then the system is programmed. Through the innovation of designers and the variety of design options, any of these pieces—the PrCB, the microchip, or the software—can carry key military mission data, and any of them may be vulnerable to a variety of threats. While each system must be analyzed to determine the vulnerabilities and threats to which it may be subject, the committee believes that printed circuit technology is of critical importance to many U.S. defensive efforts.

## FINDINGS

A number of facts were reported to the committee in the course of its data-gathering efforts. Some of these are included in the findings presented in this section.

Defense acquisition is increasingly focused on electronics as the backbone for a transformed military force. Printed circuit boards are used in almost every new military system and in a great number of critical legacy systems; the boards range widely in size, complexity, and importance. However, decisions to "make" or "buy" printed circuit boards have tipped toward "buy" over the past dozen years. This trend includes DoD itself, system integrators (also known as prime contractors, or primes), and a number of component manufacturers. For a variety of reasons, over the same past dozen years, DoD has reduced its oversight of subsystem sourcing for electronics.

The electronics industry itself has changed dramatically in the past 5 years. Key electronics-manufacturing knowledge and capabilities are migrating away from the United States and its key allies. Industry demographics worldwide have changed. In the recent past, the printed circuit board industry was made up of many medium-size, distributed businesses; it is now characterized by a few large companies and a moderate number of smaller businesses.

Certain industry demographics in the United States have also changed. Estimates show that there currently are 21 manufacturers of rigid PrCBs in the United States and 8 for flex boards for DoD. The printed circuit industry was formerly diverse; today, it is primarily made up of small businesses focused on niche, high-end industries. It is generally accepted that the majority of the production of and markets for PrCBs have gone worldwide in the past decade; as this outsourcing has progressed, the evidence shows that the level of research and development (R&D) in the industry has dropped. What smaller amount of R&D currently done is focused on process efficiencies and not on developing new products. This trend, not unique to the printed circuit board industry, is indicative of a new innovation cycle that is evolving and is no longer linear. Small, focused innovation cycles appear to exist within many steps in the supply chain. This trend makes innovation more difficult to capture, direct, tie together, apply, track, and measure.

Key advances in electronics technology are increasingly focused (to the point of exclusion) on consumer electronics. This change is driven by pressure for higher returns on investment. Competition in such high-volume manufacturing applications as personal computers, cellular telephones, and network equipment has resulted in shorter life cycles (of 2 years or less) and increased manufacturing in the Asia-Pacific region. Since the expected product life is short, requirements from underlying PrCB technology have moved from reliability and durability to performance, configuration, and cost.

In contrast with the high-growth, high-volume consumer markets, most defense requirements are for low-volume production, with highly specialized component design and life cycles that can be more than 15 years. For such components, reliability and durability are far more important than in most commercial industries. The medical electronics industry offers the best parallel, with similar low-volume/high-mix and life-cycle requirements.

A low-volume/high-mix requirement for electronics products generally leads to higher engineering and design costs. This mix also mandates premium manufacturing pricing compared with that for high-volume products such as cellular telephones. Thus, DoD will not be able to take full advantage of commercial off-the-shelf (COTS) parts and will always pay a heavy premium for custom, low-volume parts with the latest processing technology. Further, companies that compete in the high-volume products area do not have economic incentives to develop and maintain small-scale state-of-the-art PrCB facilities that can manage the DoD's volume requirements or incorporate improved manufacturing materials or processes.

The assumed patterns of technology development and technology transfer on which the DoD's policies are based are no longer valid. There is currently little incentive for large defense contractors to invest in research and development to compete for and win government contracts. This change is partly due to the pyramid of mergers and acquisitions in the defense sector, which has resulted in only a handful of companies. This lessening of competitive pressures has reduced the drive to spend internal R&D funds to improve performance. A more insidious cause may be the pressure across the government to cut costs. This pressure has resulted in a higher and higher aversion to any risk, and so new capabilities are not specified. Taken to an extreme, this could eventually lead to a stagnation in innovation of any sort.

Some agencies in the intelligence community, as an example, want access to the most cutting-edge technology. Other agencies are more risk-averse; they need to make sure that such systems as air traffic control work. However, system designers for DoD systems want access to all technology for R&D, even though they may not be able to use it immediately; one reason for wanting such access, of course, is to avoid technology surprise. This poses a process challenge to DoD to incentivize understanding and discovery of new technologies and innovation. A technology may not be critical today but strategically may be in the future. These different timescales have different criteria, different results, and require different strategies.

Finally, much speculation exists as to the role of the United States in a truly globalized economy. If in the future the United States must compete for key economic resources such as financing, oil, and electronic components, DoD supplies may be affected. For example, if policy differences arise, other countries may be able to slow down or limit access to critical DoD components that may include advanced PrCBs.

## CONSIDERATIONS

The committee identified a number of changing factors that affect the cost, availability, and quality of printed circuit boards for military applications. Among these factors are the following:

- *The variations in application, size, materials, and expected life of PrCBs cover a substantial range.* At one end of this scale are commodity products, such as the small, flex PrCBs used in cellular telephones. At the other end are products so unique that only a captive manufacturer is capable and willing to produce them.
- *Many military applications require very quick turnaround on design and production,* and the need to surge (and fade) in response to demand can be unavoidable. At one end of this scale are routine products that are ordered on a regular basis and have a relatively stable configuration, and whose production must be increased rapidly. At the other end are new products that are needed so quickly and suddenly that only a captive manufacturer may be able and willing to produce them.
- *The variations in layers, processes, and designs of PrCBs can seem limitless.* While more standard products may be easy to test for functionality, some military demands on PrCBs may cause them to fail or to perform in wholly untestable ways. The consequences either of failure or of unpredictable behavior could be of such import that only a trusted manufacturer can produce these.
- *The quality of consumer electronics can differ vastly from that of military components.* PrCBs that are used in throw-away products such as toys and cellular telephones are not likely to work in defense systems designed to be reliable and durable for up to 30 years in extreme environments. Much of today's R&D is focused on these high-volume, short-life applications.
- *The security of the product throughout the supply chain, including protection from tampering, is important to national security.* In some cases, ensuring this security may require the use of classified production facilities.
- *Compliance with environmental regulations can be problematic for defense procurement.* While military applications may be exempt from the law, many manufacturers cannot operate effectively under dual standards. As the manufacturing industry becomes more globalized, the predominance of global standards and the public pressure to maintain control of the life cycle of products and by-products across all of commercial and defense production may become dominant issues.

Dealing with any of these factors will add cost to a product. Dealing with all of them, every day, is the lot of today's DoD acquisition official.

## Manufacturing and Globalization

On top of the factors listed above, the better-known factors must also be considered—for example, delivery schedules, compliance with existing regulations (such as International Traffic and Arms Regulations [ITAR] and the Buy American Act of 1933), and compatibility with other systems. Today's acquisition officials have never been squeezed in so many directions at once. Certainly they must think beyond the "little m" view of manufacturing, generally understood to be direct production, design, and process technology. They must take a broader view of manufacturing that includes process development, supply-chain management, quality, and workforce issues. In today's business environment, an acquisition official will also consider a supplier in terms of its approach to e-business, product distribution, and strategic planning. There is currently no premium or price increment tied to the critical nature of a component, nor to the vulnerabilities or threats that may affect it. The overriding directive remains the low bid. This approach needs to be changed.

It is becoming more apparent in the trend toward globalization that DoD acquisition as an entity must take the very broad view into account. By this, the committee means the importance to the nation of manufacturing and innovation in the U.S. economy. This broad view encompasses international trade policies, geopolitics, offsets and tariffs, workforce pay differentials, the cost of insurance, the cost of environmental compliance, and many other overarching factors. These are very difficult factors to incorporate into acquisition decisions, and yet they all matter greatly.

In this much wider view, U.S. manufacturing capability directly reflects the nation's capacity for innovation, its economic development, and its focus on education and training. An important consideration is the potential futility of fostering successful innovation and then finding no U.S.

manufacturing base capable of utilizing the new ideas. Such factors will ultimately have a substantial effect on the diversity of people and thought in the United States and on our ability to prosper as a nation.

## The Separation of Innovation and Manufacturing

During the committee's deliberations, the question arose as to whether the U.S. military will be able to support the development of or even access to emerging circuit board technology if manufacturing is no longer done in the United States.

The changes in industry demographics have led to changes in the real and potential supply chain for U.S. military applications. This supply chain includes research and development. The way that outsourcing has occurred has led to a real reduction in worldwide R&D in PrCBs and to a lack of new technology capability in the United States.

The committee concluded that a more standardized framework is needed to address this challenge. Because PrCBs are extremely diverse as a technology and because their applications are equally diverse, this challenge does not have a simple solution.

In contemplating the situation, the committee observed two faces of the issue: one is that the United States will always buy the best, most cost-effective technology for the warfighter; the other is that all tax dollars should substantially benefit the taxpayer, so this country will always buy American. Clearly, both imperatives cannot be met.

A number of issues come into this debate. For example, if only critical components should be granted waivers under the Buy American Act, what is a critical component? One definition says that it is a product or service that substantially affects the warfighter. Another definition of "critical" says that it is a product or service requirement that cannot be met using conventional acquisition practices or solution paths. Yet another definition says that the critical component is of such complexity that it cannot be adequately tested using normal means in a timely manner and must therefore be procured from a trusted source.

The trusted source issue is also fraught with competing complications. Many American companies expect to be considered trusted sources simply because they are on U.S. soil. Clearly, factors such as foreign ownership, the employment of foreign nationals, and the potential for a change toward more foreign ownership, employment of more foreign workers, or even moving production or research to a foreign location, must be considered. According to an interesting alternative definition, a trusted source is one that delivers a product or service that does not necessarily need to be tested upon receipt (such as a commodity—for example, fuel from ExxonMobil). The concept of a trusted source becomes muddier upon considering the increasing move toward performance-based contracting with no process or component oversight, as well as the increasing reliance that the military is putting on fewer and fewer prime contractors. Certainly one option is to rely on the prime contractor to test and guarantee a component's performance—at least during the first round of the planned lifetime of a system.

Finally, the issue of a captive source is also difficult. Fewer and fewer processes are wholly defense-unique, and it is more and more difficult to defend keeping enhanced commercial processes under ITAR jurisdiction. A long-time push by DoD to eliminate captive sources has closed many DoD facilities and continues to threaten small shops that have filled that need.

In the course of this discussion, the committee observed that embracing only "buy the best" or only "benefit the taxpayer" is an undesirable position; it is neither realistic nor practical to lean all the way to one side or the other. Clearly, better technology is preferable, and it should be procured where it is available. However, it is the opinion of the committee that the United States is a nation of innovators. With adequate incentives, U.S. companies will produce better or equivalent technology at least part of the time. Buying American has a number of benefits that are not financial. These include improved access during wartime, easier establishment of trusted sources, and the ability to leverage other U.S.-based technology and education.

A final observation is that when the number of suppliers for any component becomes small enough, a buyer will begin to treat this small group as trusted sources and as captive sources. At that point, the responsibility as a buyer is to sustain these sources. In such a relationship, the responsibility of the source, clearly, is to stay in business. In a business sense, the implications of trusted sourcing mean that the buyer will buy no matter what, that the buyer and seller are partnered in a mutually beneficial relationship, and that neither party will take action inconsistent with that trust—even if it means higher cost for the buyer or lower return on investment for the manufacturer. A trusted supplier can be trusted to do business in a sufficiently observable, transparent, and controlled manner that unacceptable outcomes are eliminated.

Finally, the committee considered what is needed to sustain a design capability for PrCBs in the United States. The corresponding question is whether such a capability is needed. While manufacturing opened the path toward outsourcing, design has quickly followed. Therefore, if the United States is willing to lose manufacturing capabilities, DoD must be prepared to lose U.S.-based design capabilities as well.

## CONCLUSIONS AND RECOMMENDATIONS

**Recommendation 1:** The Department of Defense should address the ongoing need for printed circuit boards (PrCBs) in legacy defense systems by continuing to use the existing manufacturing capability that is resident at the Naval Surface Warfare Center, Crane Division (Indiana) and at Warner Robins Air Logistics Center (Georgia), as well as contractors currently providing legacy PrCB support.

In the short term, the captive manufacturing capability in place should continue to be used. Some very competent capability exists in a variety of places, such as (1) military facilities, including laboratories with limited production capabilities at the Warner Robins Air Logistics Center in Georgia and the Naval Surface Warfare Center, Crane Division, in Indiana; (2) small shops and boutique contractors; and (3) some defense prime contractors and their major subcontractors. This combination is currently adequate but needs to be protected; if any of these elements is shut down, action should be taken to ensure that the other elements can address the government's need. Geographic considerations should be taken into account; for example, a harsh weather event in Indiana or Georgia could have the effect of shutting down the sole source for some components for an extended period.

In addition, this network should have redundant capabilities to address needs across all services. And the network could be expanded, if desired, to encompass the needs of other government agencies, such as the intelligence community, NASA, the U.S. Postal Service, and the Air Traffic Control System—all of which have similar legacy needs. Many older commercial and municipal infrastructures are also potential users, such as older supervisory control and data acquisition (SCADA) systems that control water, electricity, and other utilities. These additional users could enhance capabilities and add robustness to such a network. Meeting some requirements for low-volume/high-mix PrCB components with life cycles that may exceed 15 years may depend on a dedicated infrastructure that tracks PrCB industry advances.

For a large number of "last resort" projects, reverse engineering is currently the plan for instances in which no commercial source can bid and no specification drawings are available. It is most desirable that the need for reverse engineering be avoided by means of provisions in all contracts to acquire design drawing and specifications in the event that they are needed. Until this can be ensured, however, the government laboratories that currently have this needed capability should be sustained. Estimates are that $1 million to $2 million per year per facility is needed to maintain the status quo. Typical equipment needed for today's production might include a laser drill, a laser trimmer, a plating line, and a direct imaging unit. Each of these pieces of equipment costs between $0.5 million and $1.5 million. Because any facility should turn over or fully refurbish its equipment every 3 to 5 years, a company with $10 million worth of equipment should be investing, on average, $3 million a year in this effort.

Capability maintenance will cost more over time, given that the number of competitors is decreasing (that is, some qualified shops are closing) and that the complexity of the technology is increasing. DoD should also consider that a periodic influx of new funding will be needed in order to get to the next generation of "legacy," as system lifetime and component use are extended past the original operational estimates.

A more permanent solution would be for DoD to provide incentives in the form of one or more joint ventures to manufacture printed circuit boards with an established firm that would locate a manufacturing facility in the United States. Because PrCB manufacturing has a much lower cost of equipment and operations compared with that for manufacturing microchips, this may be a cost-effective solution for PrCBs, whereas it would be unworkable for chips. Such a joint venture facility could also supply all classified or highly specialized boards and might also supply PrCBs to other specialized industries, such as those involving biomedical devices, law enforcement, or aerospace.

**Recommendation 2:** The Department of Defense should develop a method to assess the materials, processes, and components for manufacture of the printed circuit boards (PrCBs) that are essential for properly functioning, secure defense systems. Such an assessment would identify what is needed to neutralize potential defense system vulnerabilities, mitigate threats to the supply chain for high-quality, trustworthy PrCBs, and thus help maintain overall military superiority. The status of potentially vulnerable materials, components, and processes identified as critical to ensuring an adequate supply of appropriate PrCBs for defense systems should then be monitored.

This method must include an assessment of a number of factors. Such an assessment might call on different groups to assess each of the following areas:

- The need for an existing PrCB component or new PrCB technology should be assessed by military planning groups, and the results used to ensure access to the technologies required to field effective defense systems.
- The vulnerability of a defense system attributable to the PrCB component will require a separate assessment of operational characteristics and performance as well as potential exposures to security risks in the supply chain. The resulting information should be used to ensure the reliability and trustworthiness of PrCBs for secure, effective defense systems.
- The threat potentially posed to overall defense capabilities by lack of access to high-quality, trusted PrCB component technology will require a more specialized assessment for understanding how best to use DoD resources to maintain and enhance the nation's security.

Ultimately, these factors must be assessed for all items that depend on PrCBs, and this can easily lead to an assessment of all procured items. However, an initial assessment could fruitfully focus on PrCBs for two reasons. First, the committee believes that a number of PrCBs will exhibit the three factors listed above; and second, the sector is well enough defined that this would be a useful place to begin.

A single organization would eventually be needed to look at all of these factors together and to find their confluences and to identify potential supply shortages. Only then can the government and the defense industrial base work together to protect critical components, to analyze mitigation solutions, and to build capabilities, with the ultimate goal of efficiently ensuring supply continuity.

**Recommendation 3:** The Department of Defense (DoD) should ensure its access to current printed circuit board (PrCB) technology by establishing a competing network of shops that can be trusted to manufacture PrCBs for secure defense systems. In addition to being competitive among themselves, these suppliers should also be globally competitive to ensure the best technology for the U.S. warfighter and should be encouraged and supported to have state-of-the-art capabilities, including the ability to manufacture PrCBs that can be used in leaded and lead-free assemblies. To maintain this network of suppliers, DoD should, if necessary for the most critical and vulnerable applications, purchase more PrCBs than are required to meet daily consumption levels in order to sustain a critical mass in the trusted manufacturing base.

To facilitate this network, DoD should consider establishing a program to purchase very low volume boards on a more-frequently-than-required—and more consistent—basis. Although these purchases might not be strictly necessary, having a supply of key components as a cushion can be useful for testing, training, or logistics management. More importantly, by providing suppliers with relatively stable production orders, the overall quality and timeliness of production might go up and costs might come down.

DoD needs to become a much better supply-chain manager. Without the crutch of military specifications, DoD should work with its prime contractors to leverage the best practices of the electronics industry. Consideration should be given to both high-volume/low-mix industries, such as consumer electronics, and low-volume/high-mix industries with similar reliability concerns, such as biomedical devices, off-road vehicles, or high-rise elevators.[1] Such a strategy might be able to draw more potential suppliers into the pool through a better system for negotiation of requirements and qualifications.

Also in the short term, a government-industry consortium should be established for small businesses in order to help make them more competitive, both to fill legacy needs and to provide stewardship of this important industry sector in the move to new technology capabilities. In this way, a small, diverse group of producers could be a sustainable source of military PrCB technology. For very urgent and low-volume needs, a portion of this network should be a program adequately funded to keep a captive capability with the ability to fabricate a wide variety of boards quickly. This capability should be redundant in order to ensure the availability of needed PrCBs.

Finally, a model for close study may be the European PrCB industry. During the volatility of the past decade, many European PrCB manufacturers have remained commercially viable and sell into the high-volume consumer market. This success has supported a strong material and equipment supply base in Europe.

**Recommendation 4:** The Department of Defense (DoD) should ensure access to new printed circuit board (PrCB) technology by expanding its role in fostering new PrCB design and manufacturing technology. DoD should sponsor aggressive, breakthrough-oriented research aimed at developing more flexible manufacturing processes for cost-effective, low-volume production of custom PrCBs. In conjunction with this effort, DoD should develop explicit mechanisms to integrate emerging commercial PrCB technologies into new defense systems, even if that means subsidizing the integration. These mechanisms should include more innovative design capabilities and improved accelerated testing methods to ensure PrCBs' lifetime quality, durability, and compliance with evolving environmental regulations for the conditions and configurations unique to DoD systems.

Because even the most capable small businesses have difficulty discovering and implementing innovations, DoD needs to take a larger role in fostering new PrCB and manufacturing technology.

For the long term, state-changing research should be initiated to develop more-flexible manufacturing processes in order to enable the manufacturing of low-volume custom boards cost-effectively on high-volume commercial lines. In addition, programs should be initiated to integrate more-innovative design capabilities to take advantage of new commercial technologies in both new and legacy systems. Additional research should be initiated to develop improved accelerated testing methods to ensure lifetime quality and durability of PrCBs. This research should focus on improved reliability and durability for the unique conditions and configurations of DoD systems.

To improve security, DoD-sponsored research should improve testing and inspection capabilities to enable detection of even the most subtle defects or other variabilities in order to ensure the stated performance of printed circuit boards. Corresponding incentives should be made for programs to integrate more innovative design into boards to improve security without sacrificing performance. Future designs might include more modular PrCB configurations that would allow newer technologies or capabilities to be easily plugged in later. An advantage of more modular design is that some elements could be procured through open solicitations, and a smaller fraction that carry proprietary or classified information would need to be manufactured by trusted sources. A modular design could allow these specialized elements to be plugged in (rather than manufactured with) the base product.

## A PATH FORWARD

If the recommendations presented above are followed, the United States will be able to compete in the world market even for those items that are embedded in circuit boards and other systems. The strategy offered here will allow the United States to participate in and foster a free market society and to be a player in the world's free market system. The committee believes that, as an unintended consequence of the growth of the global PrCB marketplace, the DoD's ability to obtain needed PrCBs, in a secure manner, could be at serious risk.

---

[1] Advanced Technology Institute. 2003. Managing Low Volume Parts. Final Report For DLA Aging Aircraft Initiative, Contract No. N001140-01-C-L622.

As insurance against unforeseen consequences of fully embracing the free market, DoD must recognize that some very specific needs will manifest periodically that will demand the highest reliability. If these needs are not met at certain periods of stress, that is, during wartime, DoD will suffer the consequences of failure. The recommendations above address what is at the heart of this issue. Put another way, the best way to kill a virus is to keep the host healthy.

The committee believes that an integrated solution is reflected in its four recommendations and provides a lower-cost, lower-risk solution to an increasingly threatening situation. The committee further believes that all four recommendations should be implemented in parallel. The essence of this report is that DoD must better understand PrCB supply-chain and technology issues and must have the ability to manage research and production.

DoD must ensure that each of the recommendations is accomplished by an independent office or certainly one granted the authority of independence. Regarding Recommendations 1 and 2, it is important that DoD receive the best information and analysis on the respective subjects. It is important to realize that bad facts make for bad decisions, and DoD must strive to understand the reliability of the information it receives. It is absolutely important that the administration (and subsequent administrations) recognize the importance of reliable data and ensure that there is an independence of operation that will allow the current defense officials to give a directive without restraint. Once the nation can trust the information from Recommendations 1 and 2, then Recommendations 3 and 4 can be addressed.

The United States must have a safety net in the form of domestic manufacturing if, for example, those identified as trusted sources turn out to be less than satisfactory during hostilities. The nation cannot leave such a possibility to chance; the stakes are far too high to marginalize this issue and accept failure. Insurance is thus needed to offset such risk.

An organic, or in-house, government manufacturing capability is needed, with the ability to surge the quantities that would be needed if trusted sources should prove unavailable or unreliable. This capability can be accomplished only if DoD has a facility that it operates and can dictate the production requirements. This facility could be contractor-owned, contractor-operated; it could be government-owned, contractor-operated; or it could be government-owned, government-operated. The facility would exist to provide a center of influence or knowledge; it could have a "warm line" in operation to facilitate rapid acceleration, if required.

Although the approach of having an in-house, government manufacturing capability advocates additional expense in the procurement process, experience has shown that regardless of the completeness or perceived efficiency of war plans existing at the start of hostilities, such an organic capability within DoD control would be invaluable in such an eventuality.

Further, such a facility should be assigned to a DoD agency with responsibility and authority to develop and maintain adequate budgets to meet DoD needs. The agency should be prepared to report to the Secretary of Defense the extent of industrial preparedness planning. Such tracking should be a constant and essential element of agency reporting. Keeping this level of visibility on the subject would protect the function from inevitable budget cuts that result from the influence of those who have not experienced or do not fully appreciate the problems of supplying the logistics of a nation going to war or of sustaining a nation at war.

# Appendixes

# Appendix A

# Committee Members

**David J. Berteau**, *Chair*, is a director in the Washington, D.C., office of Clark and Weinstock. He consults with clients on federal government management, acquisition, and procurement and specializes in defense and homeland security issues. Mr. Berteau has more than 30 years of experience in public- and private-sector management. During that period, he spent 15 years in senior defense management positions, including service as Acting Assistant Secretary of Defense for Production and Logistics. He is also currently an adjunct professor at Syracuse University's Maxwell School. Mr. Berteau is a fellow of the National Academy of Public Administration and serves on the board of directors for the Procurement Roundtable. He is a frequent speaker at conferences on a wide variety of public policy issues and holds an MPA from the LBJ School of Public Affairs from the University of Texas at Austin.

**Katharine G. Frase** is vice president of Worldwide Packaging and Test of the IBM Microelectronics Division. She is responsible for all process development, design and modeling methodology, and testing and manufacturing for organic and ceramic chip packaging for IBM. Her research interests include mechanical properties and structural interactions in composites, high-temperature superconductors, solid electrolytes (fast ionic conductors), ceramic powder synthetic methods, and ceramic packaging. She chaired an IBM/National Research Council (NRC) workshop on lead solder reduction actions, and in 1998 served as the packaging assurance manager for IBM worldwide. Dr. Frase received an AB in chemistry from Bryn Mawr College and a PhD in materials science and engineering from the University of Pennsylvania. Dr. Frase is an ex officio member of the NRC's Board on Assessment of NIST (National Institute of Standards and Technology) Programs and is currently the vice chair of the Panel on Materials Science and Engineering. She is also a member of the National Materials Advisory Board.

**Charles R. Henry** has more than 32 years of senior leadership within the executive branch of the federal government and 11 years of experience as a senior executive working in civilian industry. General Henry's strategic vision revolutionized Department of Defense acquisition and procurement. He created the Defense Contract Management Command and served as its first commander. He also established the Army's Competition Advocate General's Office and served as the senior acquisitions executive for the Defense Logistics Agency during Operations Desert Shield and Desert Storm. General Henry has had a distinguished military career, having served in the Vietnam War, and has received the Defense Distinguished Service Medal, two Distinguished Service Medals, the Defense Superior Service Medal, Legion of Merit, and a Bronze Star. He currently serves on the Procurement Roundtable.

**Joseph LaDou** is director of the International Center for Occupational Medicine at the University of California, San Francisco (UCSF). His current research interests include microelectronics industry health and safety and international migration of hazardous industries. His study of the latter has led to efforts to control occupational and environmental hazards. He was in charge of the postgraduate Occupational and Environmental Medicine Program at UCSF for more than 25 years, during which 2,000 occupational physicians attended his courses. Dr. LaDou has trained more than 300 physicians who have returned to developing countries. He is editor-in-chief of the *International Journal of Occupational and Environmental Health*, a medical journal delivered to more than 100 countries. His textbook *Current Occupational and Environmental Medicine* is now in its third edition. Dr. LaDou is also the author of numerous articles in books and scientific journals discussing the health and safety problems of the high-technology electronics industry.

**Kathy Nargi-Toth**, a global business director for Technic, Inc., in Cranston, Rhode Island, is responsible for the global marketing of the company's electronics fabrication materials. Prior to her position at Technic, Ms. Nargi-Toth worked as a technical marketing manager for a large multinational chemical and laminate supply company. She also spent 6 years in a global technical marketing position that was heavily focused on the developing printed circuit infrastructure in China and the rapid technically advancing infrastructure in Taiwan, Korea, Thailand, Malaysia, and Singapore. Ms. Nargi-Toth began her career in the research and development laboratory developing electroless nickel and solder removal products for the electronics industry. Early in her career, she also worked for 8 years as a multilayer process engineer and then as engineering manager for a major supplier of multilayer printed circuit boards to the military and computer industry. Ms. Nargi-Toth currently serves on the Suppliers Council Steering Committee and chairs the Supplier's Council PCB Leadership Meeting Subcommittee. In 2000, she received the PC Fab and Atomic Giant Award recognizing her as being one of the 10 most influential individuals in the electronics industry for that year. This recognition was awarded for her activities related to microvia and high-density interconnect technology advancements in Europe and Asia. She is on the *PC Fab Magazine*'s editorial review board and has served on the board of directors of the Interconnect Technology Research Institute (ITRI).

**Angelo M. Ninivaggi, Jr.**, serves as the director of legal services for Plexus Corporation, an industry-leading electronics contract manufacturing services company. He has both broad and specific familiarity with business and management practices, including the legal issues surrounding product liability, audit responsibility, and supply-chain management. He was executive vice president, general counsel, and corporate secretary for MCMS, another contract manufacturing firm, until MCMS was acquired by Plexus. From March 1996 until joining MCMS in February 1998, he served as corporate counsel with Micron Electronics, Inc. Prior to his employment with Micron Electronics, he worked as an associate with the law firm of Weil, Gotshal and Manges in New York. Mr. Ninivaggi holds a BA in economics from Columbia University and both an MBA in finance and a JD from Fordham University. He is a member of the NRC Board on Manufacturing and Engineering Design.

**Michael G. Pecht** is founder and director of the Computer Aided Life Cycle Engineering (CALCE) Electronic Products and Systems Center at the University of Maryland and Chair Professor. The center is dedicated to providing a knowledge and resource base to support the development of competitive electronic components, products, and systems. Dr. Pecht has also consulted for more than 80 major international electronics companies, providing expertise in strategic planning, design, testing, intellectual property, and risk assessment of electronics products. He is a professional engineer, a fellow of the Institute of Electrical and Electronics Engineers (IEEE), a fellow of the American Society of Mechanical Engineers, and a Westinghouse fellow. Dr. Pecht has written 13 books on electronic product development and is chief editor for *Microelectronics Reliability* and an associate editor for the *IEEE Transactions on Advanced Packaging*. He has a BS in acoustics, an MS in electrical engineering, and an MS and PhD in engineering mechanics from the University of Wisconsin. Dr. Pecht has previously served on the NRC's Graduate Panel on Engineering.

**E. Jennings Taylor** is founder, chief technical officer, and intellectual property director of Faraday Technology, Inc.; the company develops, patents, and commercializes electrochemical technologies. He is also founder and member of Faraday Technology Marketing Group LLC, an intellectual property holding company established to provide financing and management expertise for intellectual assets. Dr. Taylor has been admitted as a registered patent agent to prepare and prosecute patent applications before the U.S. Patent and Trademark Office. He has held appointments with the National Science Board, the National Science Foundation, the Edison Materials Technology Center in Ohio, the Electrochemical Society, and the American Electroplaters and Surface Finishers Society. He holds a PhD in materials science from the University of Virginia.

**Richard H. Van Atta** is a senior research analyst at the Institute for Defense Analyses (IDA), focusing on the technological needs and interests of the United States as they affect both national and economic security. At IDA, Dr. Van Atta has worked on numerous issues including microelectronics industry assessments in support of the Defense Science Board and the Office of the Under Secretary of Defense for Acquisition and Technology. From 1993 to 1998, he was an official in the Department of Defense (on temporary assignment from IDA), first as Special Assistant for Dual Use Technology Policy, then as Assistant Deputy Under Secretary for Dual Use and Commercial Programs. When in the Office of the Secretary of Defense, he developed policy options and conducted assessments related to the dual-use concept for developing and acquiring defense systems and for promoting the industrial and technological base needed for meeting defense needs. He also played a major role in authoring the Department of Defense's (DoD's) Dual Use Technology Strategy and conceived and managed the Commercial Technology Insertion Program. For his service, Dr. Van Atta was presented with DoD's Award for Outstanding Achievement by Secretary of Defense William Perry.

**Alfonso Velosa III** is the associate director of market and business strategies at Gartner, Inc. In this position, he manages and contributes to a variety of custom market-assessment and strategy projects in the global technology marketplace, focusing on the semiconductor, telecommunications, and manufacturing arenas. His work experience covers a broad range of areas in the technology marketplace, including strategic planning, project and program management, supply-chain management, contract negotiations, profit and loss analysis, product management support, and semiconductor research. Prior to joining Gartner, Mr. Velosa was a commodity specialist for Intel. In this role, he managed Intel's motherboard and server products from concept through supplier selection through design through production. In addition, he managed the supply chain and negotiated overall relationships and terms with suppliers. Previously, he provided project and program management services to NASA in Washington, D.C., and in Cleveland, Ohio, culminating in a project management role for a semiconductor diffusion project that flew on the space shuttle in 1997. He holds a BS in materials science engineering from Columbia University, an MS in materials science engineering from Rensselaer Polytechnic Institute, and another master's degree in international management from Thunderbird, The American Graduate School of International Management, in Glendale, Arizona. He is a member of the NRC Board on Manufacturing and Engineering Design.

**Dennis F. Wilkie** is senior vice president in the Management Consulting Division of Compass Group, Ltd., which provides consulting in strategy, marketing, product development, and operations to Senior Management of Companies. He is an accomplished senior executive with an unusual mix of automotive manufacturer and component supplier perspectives through operating assignments in Ford Motor Company and Motorola, Inc. Dr. Wilkie has a strong technical background combined with bottom-line operational experience as a general manager. He has a demonstrated record of leadership in quality management, people skills, and cross-cultural management. He was elected to the National Academy of Engineering in 2000 for the application of electronics and systems engineering technology to vehicular systems.

# Appendix B

# Selected Abbreviations and Acronyms

| | |
|---|---|
| BFL | beam forming lens |
| BFR | brominated flame retardant |
| COTS | commercial off-the-shelf |
| DARPA | Defense Advanced Research Projects Agency |
| DIB | defense industrial base |
| DLA | Defense Logistics Agency |
| DMEA | Defense Microelectronics Activity |
| DoD | Department of Defense |
| DSTL | Developing Science and Technologies List |
| DUSD | Deputy Under Secretary of Defense |
| EAR | Export Administration Regulations |
| E/CIT | Emerging Critical Interconnect Technology |
| EPA | Environmental Protection Agency |
| EU | European Union |
| FSC | Federal Supply Class |
| FPGA | field programmable gate array |
| IBM | International Business Machines |
| IC | integrated circuit |
| iNEMI | International Electronics Manufacturing Initiative |
| IP | Industrial Policy |
| IPC | Association Connecting Electronics Industries |
| ITAR | International Traffic in Arms Regulations |
| ITRI | Interconnect Technology Research Institute |
| LCP | liquid crystal polymer |
| MCTL | Militarily Critical Technologies List |
| MEMS | microelectromechanical systems |
| MIL-PRF | military performance specification |
| NASA | National Aeronautics and Space Administration |
| NEMI | National Electronics Manufacturing Initiative |
| NIIN | National Item Identification Number |

| NIST | National Institute of Standards and Technology |
| NRC | National Research Council |
| NSN | National Stock Number |
| OEM | original equipment manufacturer |
| PLC | programmable logic controlled |
| PMTEC | Printed Wiring Board Manufacturing Technology Center |
| PrCB | printed circuit board |
| PWB | printed wiring board |
| R&D | research and development |
| RCRA | Resource Conservation and Recovery Act of 1976 |
| REACH | Regulation, Evaluation, and Authorization of Chemicals |
| RF | radio frequency |
| RoHS | restriction of hazardous substances |
| SAIC | Science Applications International Corporation |
| SEMATECH | SEmiconductor MAnufacturing TECHnology |
| SME | small and medium enterprises |
| SMOBC | solder mask over bare copper |
| USML | U.S. Munitions List |
| WEEE | Waste Electrical and Electronic Equipment |

# Appendix C

# Agenda of the
# Workshop on Manufacturing Trends for
# Printed Circuit Technology

Day 1: Monday, December 13, 2004

| | | |
|---|---|---|
| 8:45 a.m. | Welcome and Introductions<br>Purpose of Meeting | David Berteau, Chair |
| 9:00 a.m. | I: State of the Industry | David Bergman, IPC<br>Frank Vargo, National Association of Manufacturers<br>Hayao Nakahara, NT Information, Ltd.<br>Harvey Miller, Fabfile |
| 10:45 a.m. | II: Military and Legacy Needs | John Shawhan, Warner Robins Air Logistics Center<br>Roger Smith, Naval Surface Warfare Center<br>James Kachmarsky, Tobyhanna Army Depot |
| 1:15 p.m. | III: Industry Perspective | Herm Reininga, Rockwell Collins<br>Richard Snogren, Coretec<br>Don Dupriest, Lockheed Martin<br>Charles Mullins, Raytheon<br>Benoit Pouliquen, Multek |
| 3:00 p.m. | IV: State of the Technology | Jack Fisher, iNEMI<br>David Fries, University of South Florida<br>Wayne Johnson, Auburn University |
| 4:30 p.m. | Roundtable discussion | |
| 5:30 p.m. | Adjourn | |

Day 2: Tuesday, December 14, 2004

| | | |
|---|---|---|
| 9:00 a.m. | V: Potential Approaches | Dick Pinto, Advisor<br>Darrel Frear, Freescale<br>Carolynn Drudik, Defense Microelectronics Activity<br>Ron Thompson, SAIC |

11:00 a.m.   Facilitated discussion

Noon   Adjourn

# Appendix D

# Workshop Attendees

Fern Abrams
IPC

Dick Alexander
Rockwell Collins

Doug Bartlett
Bartlett Manufacturing

David Bergman
IPC

David J. Berteau
Clark and Weinstock

Jamie Blackwell
Naval Surface Warfare Center, Crane Division

Dennis Chamot
National Materials Advisory Board

Carolynn Drudik
DMEA

Don Dupriest
Lockheed

Robert Ek
Clark and Weinstock

Jack Fisher
iNEMI

Katharine G. Frase
IBM

Darrel Frear
Freescale

David Fries
University of South Florida

Steve Gootee
SAIC

Charles R. Henry
U.S. Army, retired

Wayne Johnson
Auburn University

James Kachmarsky
Tobyhanna Army Depot

John Kania
IPC

Joseph LaDou
University of California

William Marck
U.S. House of Representatives Staff, Armed
    Services Committee

Toni Marechaux
Board on Manufacturing and Engineering
    Design

Harvey Miller
Fabfile

Charles Mullins
Raytheon

Hayao Nakahara
NT Information, Ltd.

Kathy Nargi-Toth
Technic, Inc.

James Passanisi
Raytheon

Michael Pecht
University of Maryland

Dick Pinto
Consultant

Benoit Pouliquen
Multek

Jean Reed
U.S. House of Representatives Staff, Armed
    Services Committee

Herm Reininga
Rockwell Collins

Chuck Romine
Office of Science and Technology Policy

John Shawhan
Warner Robins Air Logistics Center

Susan Skemp
Office of Science and Technology Policy

Roger Smith
Naval Surface Warfare Center, Crane Division

Richard Snogren
Coretec

George Steimle
University of South Florida

Dave Sullivan
Rockwell Collins

Bob Sweet
U.S. House of Representatives Staff, Education
    and the Workforce Committee

E. Jennings Taylor
Faraday Technology, Inc.

Ron Thompson
SAIC

Laura Toth
Board on Manufacturing and Engineering
    Design

Richard H. Van Atta
Institute for Defense Analyses

Frank Vargo
National Association of Manufacturers

Alfonso Velosa
Gartner, Inc.

Marta Vornbrock
Board on Manufacturing and Engineering
    Design

Al Wavering
National Institute of Standards and Technology

Rick Weeks
Warner Robins Air Logistics Center

Dennis F. Wilkie
Consultant

# Appendix E

# Lead-Free Electronics

Lead is a common component in electronics manufacturing, adding functionality to a variety of solders, capacitors, glasses, and paints. For as long as lead has been used, however, the hazards of lead mining, smelting, industrial use, and recycling have also been known.

The Department of Defense (DoD) has put forth a number of technical reasons why defense acquisition need not fully comply with European Union (EU) regulations on the use of lead in electronics products.[1] While many manufacturers in the printed circuit board (PrCB) industry are expected to be able to meet the standards set in the European Union, and in California, some board manufacturers are seeking exemptions from the European Union so that compliance can be delayed.[2]

## HEALTH EFFECTS OF LEAD

Toxic lead exposure is the most significant and prevalent disease of environmental origin in the world today. Despite all that is known regarding the hazards of lead exposure for children, it has taken over a century for primary prevention to be adopted in the most highly developed countries. The developing world is woefully behind in the development of programs to protect children from lead poisoning. The demonstrable success and societal benefits of preventing lead exposure are unarguable.

Irrefutable evidence associates lead at different exposure levels with a wide spectrum of health and social effects, including mild intellectual impairment, hyperactivity, shortened concentration span, poor school performance, violent or aggressive behavior, and hearing loss. Lead has an impact on virtually all organ systems, including the heart, brain, liver, kidneys, and circulatory system, resulting in coma and death in severe cases.

---

[1] Restriction of hazardous substances (RoHS), lead-free legislation, or, more accurately, "Directive 2002/95/EC on the restriction of the use of certain hazardous substances in electrical and electronic equipment," will be enforced throughout the European Community beginning July 1, 2006.

[2] Fern Abrams. 2005. Lead free electronics: Should the military be concerned? Presentation at the Diminishing Manufacturing Sources and Materials Shortages (DMSMS) Conference, Nashville, Tenn., April 14.

Lead's neurotoxicity has long been recognized in industrial workers in lead processing and production and in the health care community.[3-11] Many studies relate increased blood pressure and hypertension in adults to elevated blood lead levels.[12-20] These conditions, in turn, increase the risk of cardiovascular disease. The effect of lead on blood pressure, a major risk factor for coronary artery disease and stroke, is seen at levels quite prevalent in the general population. Lead also

---

[3] M. Payton, K.M. Riggs, A. Spiro, S.T. Weiss, and H. Hu. 1998. Relations of bone and blood lead to cognitive function: The VA normative aging study. Neurotox Teratol. 20:19-27.

[4] D. Rhodes, A. Spiro, A. Aro, and H. Hu. 2003. Relationship of bone and BLLs to psychiatric symptoms: The VA normative aging study. J. Occup. Environ. Med. 45:1144-1451.

[5] B.S. Schwartz, B.-K. Lee, G.S. Lee, W.F. Stewart, S.S. Lee, K.Y. Hwang, K.D. Ahn, Y.B. Kim, K.I. Bolla, D. Simon, P.J. Parsons, and A.C. Todd. 2001. Association of blood lead, dimercaptosuccinic acid-chelatable lead, and tibia lead with neurobehavioral test scores in South Korean lead workers. American Journal of Epidemiology 53:453-464.

[6] B.S. Schwartz, W.F. Stewart, K.I. Bolla, M.S. Simon, K. Bandeen-Roche, B. Gordon, J.M. Links, and A.C. Todd. 2000. Past adult lead exposure is associated with longitudinal decline in cognitive function. Neurology 55:1144-1150.

[7] B.S. Schwartz, B.-K. Lee, K. Bandeen-Roche, W.F. Stewart, K.I. Bolla, J. Links, V. Weaver, and A. Todd. 2005. Occupational lead exposure and longitudinal decline in neurobehavioral test scores. Epidemiology 16:106-113.

[8] W.F. Stewart, B.S. Schwartz, D. Simon, K.I. Bolla, A.C. Todd, and J. Links. 1999. Neurobehavioral function and tibial and chelatable lead levels in 543 former organolead workers. Neurology 52:1610-1617.

[9] R.O. Wright, S.W. Tsaih, J. Schwartz, A. Spiro, K. MacDonald, S.T. Weiss, and H. Hu. 2003. Independent and modifying effects of lead biomarkers on minimental status exam scores in elderly men: The normative aging study. Epidemiology 14:713-718.

[10] N. Fiedler, C. Weisel, R. Lynch, K. Kelly-McNeil, R. Wedeen, K. Jones, I. Udasin, P. Ohman-Strickland, and M. Gochfeld. 2003. Cognitive effects of chronic exposure to lead and solvents. American Journal of Industrial Medicine 44:413-423.

[11] M.G. Weisskopf, R.O. Wright, J. Schwartz, A. Spiro, D. Sparrow, A. Aro, and H. Hu. 2004. Cumulative lead exposure and prospective change in cognition among elderly men: The normative aging study. American Journal of Epidemiology 160:1184-1193.

[12] Y. Cheng, J. Schwartz, D. Sparrow, A. Aro, S.T. Weiss, and H. Hu. 2001. Bone lead and BLLs in relation to baseline blood pressure and the prospective development of hypertension: The normative aging study. American Journal of Epidemiology 153:164-171.

[13] B.S. Glenn, W.F. Stewart, J.M Links, A.C. Todd, and B.S. Schwartz. 2003. The longitudinal association of lead with blood pressure. Epidemiology 14:30-36.

[14] B.S. Glenn, W.F. Stewart, B.S. Schwartz, and J. Bressler. 2001. Relation of alleles of the sodium-potassium adenosine triphosphatase alpha 2 gene with blood pressure and lead exposure. American Journal of Epidemiology 153:537-545.

[15] H. Hu, A. Aro, M. Payton, S. Korrick, D. Sparrow, S.T. Weiss, and A. Rotnitzky. 1996. The relationship of bone and blood lead to hypertension. JAMA 275:1171-1176.

[16] S.A. Korrick, D.J. Hunter, A. Rotnitzky, H. Hu, and F.E. Speizer. 1999. Lead and hypertension in a sample of middle-aged women. American Journal of Public Health 89:330-335.

[17] B.K. Lee, G.S. Lee, W.F. Stewart, K.D. Ahn, D. Simon, K.T. Kelsey, A.C. Todd, and B.S. Schwartz. 2001. Associations of blood pressure and hypertension with lead dose measures and polymorphisms in the vitamin D receptor and d-aminolevulinic acid dehydratase genes. Environmental Health Perspectives 109:383-389.

[18] D. Nash, L. Magder, M. Lustberg, R. Sherwin, R. Rubin, R. Kaufmann, and E. Silbergeld. 2003. Blood lead, blood pressure, and hypertension in perimenopausal and postmenopausal women. JAMA 289:1523-1531.

[19] S.J. Rothenberg, V. Kondrashov, M. Manalo, J. Jiang, R. Cuellar, M. Garcia, B. Reynoso, S. Reyes, M. Diaz, and A.C. Todd. 2002. Increases of hypertension and blood pressure during pregnancy with increased bone lead. American Journal of Epidemiology 156:1079-1087.

[20] J. Schwartz. 1988. The relationship between blood lead and blood pressure in NHANES II survey. Environmental Health Perspectives 78:15-22.

damages the kidneys and causes anemia.[21-25] Lead has a number of untoward effects on reproductive health[26-31] and may be a cause of cataracts.[32] Recent epidemiological and experimental work confirms that inorganic lead compounds are associated with increased risks of cancer.[33]

Various chelating agents have been used to treat lead poisoning. Unfortunately, because many people think that a treatment for lead poisoning exists, they see no further reason to limit lead exposure. There is no conclusive evidence that chelation improves therapeutic outcome in patients with lead poisoning.[34] Although chelation reduces blood lead levels and increases excretion of lead in the urine, there is very little evidence that it prevents or reverses the damage resulting from lead exposure.[35] Moreover, chelation does not have a beneficial effect on growth and may even have an adverse effect.[36]

## Global Lead Exposure

The public health problem of environmental lead exposure has been widely investigated in developed countries such as the United States, where actions taken have led to significant reductions in blood lead concentrations in children. In contrast, little has been done regarding lead poisoning in developing countries, particularly in African countries, despite evidence of widespread and excessive lead

[21] R. Kim, A. Rotnitzky, D. Sparrow, S.T. Weiss, C. Wager, and H. Hu. 1996. A longitudinal study of low-level lead exposure and impairment of renal function: The normative aging study. JAMA 275:1177-1181.

[22] J.L. Lin, D.T. Lin-Tan, K.H. Hsu, and C.C. Yu. 2003. Environmental lead exposure and progression of chronic renal diseases in patients without diabetes. New England Journal of Medicine 348:277-286.

[23] M. Payton, H. Hu, D. Sparrow, and S.T. Weiss. 1994. Low-level lead exposure and renal function in the normative aging study. American Journal of Epidemiology 140:821-829.

[24] J.A. Staessen, R.R. Lauwerys, J.P. Buchet, C.J. Bulpitt, D. Rondia, Y. Vanrenterghem, and A. Amery. 1992. Impairment of renal function with increasing blood lead concentrations in the general population: The Cadmibel study group. New England Journal of Medicine 16:151-156.

[25] S.-W. Tsaih, S. Korrick, J. Schwartz, C. Amarasiriwardena, A. Aro, D. Sparrow, H. Hu. 2004. Lead, diabetes, hypertension, and renal function: The normative aging study. Environmental Health Perspectives 112:1178-1182.

[26] A. Gomaa, H. Hu, D. Bellinger, J. Schwartz, S. Tsaih, T. Gonzalez-Cossio, L. Schnaas, K. Peterson, A. Aro, and M. Hernandez-Avila. 2002. Maternal bone lead as an independent risk factor for fetal neurotoxicity: A prospective study. Pediatrics 110:110-118.

[27] T. Gonzalez-Cossio, K.E. Peterson, L. Sanin, S.E. Fishbein, E. Palazuelos, A. Aro, M. Hernandez-Avila, and H. Hu. 1997. Decrease in birth weight in relation to maternal bone lead burden. Pediatrics 100:856-862.

[28] B.L. Gulson, C.W. Jameson, K.R. Mahaffey, K.J. Mizon, N. Patison, A. Law, M.J. Korsch, and M.A. Salter. 1998. New findings on sources and biokinetics of lead in human breast milk: Bone lead can target both nursing infant and fetus. Environmental Health Perspectives 106: 667-674.

[29] B.L. Gulson, K.J. Mizon, M.J. Korsch, J.M. Palmer, and J.B. Donnelly. 2003. Mobilization of lead from human bone tissue during pregnancy and lactation: A summary of long-term research. Sci. Total Environ. 303:79-104.

[30] M. Hernandez-Avila, K.E. Peterson, T. Gonzalez-Cossio, L.H. Sanin, A. Aro, L. Schnaas, and H. Hu. 2002. Effect of maternal bone lead on length and head circumference at birth. Archives of Environmental Health 57:482-488.

[31] L.H. Sanin, T. Gonzalez-Cossio, I. Romieu, K.E. Peterson, S. Ruz, E. Palazuelos, M. Hernandez-Avila, and H. Hu. 2001. Effect of maternal lead burden on infant weight and weight gain at one month of age among breastfed infants. Pediatrics 107:1016-1023.

[32] D.A. Schaumberg, F. Mendes, M. Balaram, M.R. Dana, D. Sparrow, and H. Hu. 2004. Accumulated lead exposure and risk of age-related cataract extraction in men: The normative aging study. JAMA 292:2750-2754.

[33] E.K. Silbergeld, M. Waalkes, and J.M. Rice. 2000. Lead as a carcinogen: Experimental evidence and mechanisms of action. Am. J. Ind. Med. 38:316-323. Review.

[34] K.N. Dietrich, J.H. Ware, M. Salganik, J. Radcliffe, W.J. Rogan, G.G. Rhoads, M.E. Fay, C.T Davoli, M.B. Denckla, R.L. Bornschein, D. Schwarz, D.W. Dockery, S. Adubato, and R.L. Jones. 2004. Treatment of lead-exposed children clinical trial group: Effect of chelation therapy on the neuropsychological and behavioral development of lead-exposed children after school entry. Pediatrics 114:19-26.

[35] W.J. Rogan, K.N. Dietrich, J.H. Ware, D.W. Dockery, M. Salganik, J. Radcliffe, R.L. Jones, N.B. Ragan, J.J. Chisolm, Jr., and G.G. Rhoads. 2000. Treatment of lead-exposed children trial group: The effect of chelation therapy with succimer on neuropsychological development in children exposed to lead. New England Journal of Medicine 344:1421-1426.

[36] K.E. Peterson, M. Salganik, C. Campbell, G.G. Rhoads, J. Rubin, O. Berger, J.H. Ware, and W. Rogan. 2004. Effect of succimer on growth of preschool children with moderate blood lead levels. Environmental Health Perspectives 112:233-237.

exposure during childhood.[37,38] Estimates of lead exposure and surveys of blood lead in the most populous developing countries, India, Indonesia, and China, indicate that most of the world's children are already at risk from the effects of environmental lead exposure.[39,40] The same can be said of the risk to children in Latin America and Central Europe.[41-44]

There has been a laudable effort in many developed countries to protect the public from lead exposure. Government regulation followed public health recommendations to remove or severely limit the amount of lead in gasoline, paint, and food containers.[45] Litigation and environmental education programs have driven most lead smelting and refining out of populous areas. Vigorous enforcement of environmental standards in one country, however, has resulted all too often in a transfer of its industrial hazards to other, poorer countries.[46] The record of decreasing lead exposure in developed countries masks a shameful record of the export of lead to developing countries, most of which are unaware of the extent of lead's toxicity and equally unaware of the long-term costs associated with environmental degradation and damage to the health of workers and community residents.

Public health agencies in many countries have responded to growing awareness of the severity and prevalence of lead poisoning by promulgating regulations to reduce exposure to lead.[47] These regulations apply to lead in products, environmental media, or the workplace and define restrictions on specific uses of lead. Public health policies have not adequately prevented lead exposure or lead poisoning. This is related in part to the complexity of limiting exposures and also to an unresolved debate over the value of screening. Despite considerable regulatory attention and voluntary changes that have occurred in reducing lead use, lead poisoning continues in developed as well as developing countries.[48]

## The Electronics Industry

The electronics industry was not regulated for its impact on the environment for many decades and has never been held accountable for the actual cost of the environmental damage that it has caused. Billions of electronics products have been discarded in every region of the world. Not until 1997 did the U.S. Environmental Protection Agency (EPA) begin to address the relationship between product and process design and environmental impact. By that time, the international pollution of the world with what has come to be known as e-waste was readily apparent.

Lead use is ubiquitous in electronics manufacturing. It is present in solder and interconnects, finishes, batteries, paints, piezoelectric ceramic devices, discrete components, sealing glasses, and

[37] L.J. Fewtrell, A. Pruss-Ustun, P. Landrigan, and J.L. Ayuso-Mateos. 2004. Estimating the global burden of disease of mild mental retardation and cardiovascular diseases from environmental lead exposure. Environ. Res. 94:120-133.

[38] D. Bellinger, H. Hu, V. Kartigeyan, T. Naveen, R. Pradeep, S. Sankar, R. Padmavathi, and B. Kalpana. 2005. A pilot study of blood lead levels and neurobehavioral function in children living in Chennai, India. Int. J. Occup. Environ. Health 11:138-143.

[39] I. Heinze, R. Gross, P. Stehle, and D. Dillon. 1998. Assessment of lead exposure in school children from Jakarta. Environmental Health Perspectives 106:499-501.

[40] M. Lacasana, I. Romieu, L.H. Sanin, E. Palazuelos, and M. Hernandez-Avila. 2000. Blood lead levels and calcium intake in Mexico City children under five years of age. Int. J. Environ. Health Res. 10:331-340.

[41] P. Factor-Litvak, G. Wasserman, J.K. Kline, and J. Graziano. 1999. The Yugoslavia prospective study of environmental lead exposure. Environmental Health Perspectives 107:9-15.

[42] L. Schnaas, S.J. Rothenberg, M.F. Flores, S. Martinez, C. Hernandez, E. Osorio, and E. Perroni. 2004. Blood lead secular trend in a cohort of children in Mexico City (1987-2002). Environmental Health Perspectives 112:1110-1115.

[43] A. Mathee, Y. von Schirnding, M. Montgomery, and H. Rollin. 2004. Lead poisoning in South African children: The hazard is at home. Rev. Environ. Health 19:347-361.

[44] I. Romieu, M. Lacasana, and R. McConnell. 1997. Lead exposure in Latin America and the Caribbean. Lead research group of the Pan-American Health Organization. Environmental Health Perspectives 105:398-405.

[45] L.R. Goldman. 1998. Linking research and policy to ensure children's environmental health. Environmental Health Perspectives 106 (Suppl 3):857-862.

[46] J. LaDou. 2003. International occupational health. Int. J. Hygiene Environ. Health 206:303-313.

[47] Organization for Economic Cooperation and Development. 1999. Review of Implementation of OECD Environment Ministerial Declaration on Risk Reduction for Lead. Paris: OECD Environmental Directorate.

[48] E.K. Silbergeld. 1995. The international dimensions of lead exposure. Int. J. Occup. Environ. Health 1:336-348.

cathode-ray-tube glass. Lead is also used as a stabilizer for plastics such as PVC (polyvinyl chloride), commonly used in cable assemblies. The early PrCB industry produced electronics products using prodigious quantities of lead and other toxic materials, systematically shipping them to every corner of the world, where, to this day, they are improperly disposed of in landfills, waterways, and incinerators.

The elimination of lead plating has been a goal of many PrCB manufacturers, in part because of strict local discharge limitations. Tin-lead is plated as an etch resist; then, on panels subsequently processed with solder-mask-over-bare-copper (SMOBC), the tin-lead coat is promptly stripped. Therefore, when the predominant SMOBC process is specified, tin-lead is easily replaced by tin as the etch resist of choice. Unfortunately, a minority of PrCBs still require tin-lead reflow, and these panels must be processed with a tin-lead etch resist, which is subsequently fused into solder. Many shops do not, for economic or other reasons, maintain both tin and tin-lead plating lines and are thus unable to employ tin-only plating on that portion of their product which is SMOBC. In short, the transition from tin-lead plating to tin-only has been slow.

As part of the EPA's Design for the Environment project, the U.S. government found that the range of water use among participants was very large, and there was evidence of wide variation in the water practices among facilities. The survey data show that the majority of the respondents are indirect dischargers (i.e., facilities that discharge process wastewaters to publicly owned treatment works). This was especially true for the small to midsize PrCB-manufacturing facilities. The regulated pollutants most often found in PrCB wastewater are copper, lead, nickel, silver, and total toxic organics. Accidental or unauthorized release of these pollutants into surface waters can harm aquatic life.[49]

The next most commonly shipped waste product is tin or tin-lead stripping solutions. Moreover, current PrCB manufacturing processes generate a variety of scrap materials containing lead, copper, tin, nickel, gold, and other metals. Most manufacturers collect these materials for recycling, disposing of them through brokers. Flux, solder dross from the hot-air-solder-level process, and other lead-bearing solutions are shipped off-site for recycling by 20 percent of the EPA survey respondents. The United States exports 50 to 80 percent of its e-waste for recycling. In many cases, manufacturers are not aware of where their brokers ship these materials.[50]

## GLOBAL ENVIRONMENTAL REGULATION

Regulatory initiatives are emerging that require the electronics industry to incorporate environmental, health, and safety considerations into design and manufacturing decisions. The electronics industry is preparing to comply with a number of restricted materials laws.

In 2003, the European Union (EU) enacted the restriction of hazardous substances (RoHS) directive that bans the use of lead, mercury, cadmium, hexavalent chromium, and certain brominated flame retardants (BFRs) in most electronics products sold in the EU market beginning July 1, 2006.[51] Both business-to-business and consumer products are covered. Although there are some exemptions to the directive's chemical restrictions, this directive, by banning the use of critical materials in electronics products sold in key world markets, may result in a significant change in the way products are designed for global sale.

The European Parliament and the European Council are considering legislation—Regulation, Evaluation, and Authorization of Chemicals (REACH)—that will require industry to prove that chemicals being sold and produced in the European Union are safe to use or handle. REACH policy will require the registration of all substances that are produced or imported into the European Union. The amount of information required for registration will be proportional to the health risks related to the chemical and its production volumes. Companies will also need to seek authorization to sell and produce problematic chemicals, such as carcinogens, mutagens, and teratogens. Toxic chemicals that persist in the

---

[49] Environmental Protection Agency. Design for the Environment. Available at http://www.epa.gov/dfe/projects/pwb/ index.htm. Accessed October 2005.

[50] Environmental Protection Agency. Design for the Environment. Available at http://www.epa.gov/dfe/projects/pwb/ index.htm. Accessed October 2005.

[51] European Union. Directive 2002/95/EC of the European Parliament and of the Council of 27 January 2003 on the Restriction of the Use of Certain Hazardous Substances in Electrical and Electronic Equipment (RoHS). Available at http://europa.eu.int/eur-lex/pri/en/oj/dat/2003/l_037/l_03720030213en00190023.pdf. Accessed October 2005.

environment or that bioaccumulate will also need authorization. The policy is slated for enactment in 2006.[52]

California recently enacted the first law in this country to establish a funding mechanism for the collection and recycling of computer monitors, laptop computers, and most television sets sold in the state. That law, the Electronic Waste Recycling Act of 2003, also contains a provision that prohibits a covered electronics device from being sold or offered for sale in California if the device is prohibited from being sold in the European Union by the RoHS directive.[53]

The electronics industry is likewise beginning to take responsibility for its products at the end of their useful life. This responsibility also forms the basis for the "take-back" legislation that is being implemented in the European Union under the Waste Electrical and Electronic Equipment (WEEE) directive, beginning in August 2005.[54] The directive encourages the design and production of electronics equipment to take into account and facilitate dismantling and recovery, in particular the reuse and recycling of electronics equipment, components, and materials necessary to protect human health and the environment.

In the European Union, since July 1, 2003, materials and components have not been allowed deliberately to contain lead, mercury, cadmium, or hexavalent chromium.[55] Lead was classified as category 1, toxic to reproduction (embryotoxic), and as a precaution, the EU classified lead chromate pigments as category 3 carcinogens.

In the United States, environmental regulation is not moving in the same direction as in Europe. In 2003, the EPA proposed revisions to the definition of solid waste that would exclude certain hazardous waste from the Resource Conservation and Recovery Act (RCRA) of 1976 if the waste is reused in a "continuous industrial process within the same generating industry." The proposal may eventually exempt all appropriately recycled materials from RCRA hazardous-waste regulations. Final action on the proposal is expected in 2006. The EPA is also considering a rule that would exempt electroplating sludge from RCRA hazardous-waste regulations if the sludge is recycled.

Workplace exposures to airborne chemicals are regulated in the United States by the Occupational Safety and Health Administration (OSHA) via the promulgation of Permissible Exposure Limits (PELs). These limits, usually defined as 8-hour time-weighted average values, are enforced as concentrations never to be exceeded. In the case of toxicants with chronic or delayed effects, the PEL is determined from toxicological or epidemiological evidence and assessment of risk of disease or material impairment of heath resulting from exposure to the PEL over an occupational lifetime of 45 years.

Industry guidelines and federal standards have failed to fully protect workers from chemical toxicity: no such guidelines or standards exist for most chemicals, many are biased toward what can easily be achieved, and many were developed long after health consequences became evident. Although exposure limits or guidelines for many large-volume chemicals have been established, federal OSHA has PELs for fewer than 500 toxic substances out of the more than 10,000 chemicals that are routinely used in industrial facilities. Additionally, more than 90 percent of the substances with established PELs have standards based on toxicological study results and case reports from 35 to 50 or more years ago: lead is among them.[56]

## TECHNICAL CHALLENGES

A number of technical challenges face the PrCB industry in this transition period. Foremost is the integration of new materials and processes coupled with the need to maintain product quality. Process

---

[52] P.D. Thacker. 2005. U.S. Companies Get Nervous about EU's REACH. Environmental Science and Technology Online, January 5. Available at http://pubs.acs.org/subscribe/journals/esthag-w/2005/jan/policy/pt_nervous.html. Accessed October 2005.

[53] California Department of Toxic Substances Control. Electronic Waste Recycling Act of 2003 (SB20). Available at http://www.dtsc.ca.gov/HazardousWaste/CRTs/SB20.html. Accessed September 2005.

[54] European Union. Directive 2002/96/EC of the European Parliament and of the Council of 27 January 2003 on Waste Electrical and Electronic Equipment (WEEE).

[55] European Union. Directive 67/548/EEC on the Classification, Packaging and Labelling of Dangerous Substances, Annex 1, as last amended by Directive 2003/32/EC (28th ATP).

[56] Historical information on permissible exposure levels is available from the Occupational Safety and Health Administration, at http://www.osha.gov. Accessed October 2005.

challenges include the higher reflow temperatures (+20°C to 40°C) for lead-free solders, which will affect soldering processes for the entire system. This change is expected to require the requalification of military suppliers and (if widely implemented) will involve potential incompatibilities with legacy systems. Current testing indicates that most products that are manufactured to be lead-free could be used with a leaded connection, but this is a difficult paradigm for military systems owners to trust.

One inherent difficulty stems from the reason that lead was introduced into coatings and solders a century ago. Pure tin metal is susceptible to spontaneous growth of filament-like structures commonly referred to as tin whiskers. The problem of tin "whiskers, needles or filaments" was first reported just after World War II. It was discovered that these whiskers could cause problems in electronic devices by breaking off and causing shorts between exposed leads. Whisker growth rates and final sizes are unpredictable, and growth can begin soon after manufacturing or may take years to initiate. Whiskers have been observed to grow with a wide range of morphologies and a wide range of diameters and lengths. Currently, a thorough understanding of the problem is lacking, and many approaches are being tested.

The cross-contamination of leaded components and no-lead components is an additional worry. Many small and some large manufacturers will not be able to maintain dual production for military and nonmilitary uses. This problem is expected to result in fewer suppliers for both systems, because some will have to choose whether or not to transition. The industry already manufactures PrCBs that can be used in leaded and no-lead systems, but this will further serve to limit qualified suppliers and will narrow supply-chain options.

Current exemptions from the restriction include high-lead solders (over 85 percent lead) because there is no viable materials substitute for these solders. Other exemptions include lead in the glass of cathode ray tubes, electronic components, and fluorescent tubes; lead in solders for servers, storage and storage array systems; lead in solders for network infrastructure equipment for switching, signaling, transmission as well as network management for telecommunication; and lead in electronic ceramic parts (e.g., piezoelectronic devices). The legislation also does not cover medical electronics and monitoring and control instrumentation.

It is important to note that all of these exemptions are subject to future legislative action and that the committee believes that the transition to no-lead systems will continue. While these exemptions may ease the design challenges inherent in any transition to a new technology, the net effect is an extension of the problem faced by designers and users. Such extensions may contribute to further cross-contamination of systems and will delay the underlying intent of the restriction, which is to eliminate hazardous waste from electronics at the beginning and end of life.

## CONCLUSIONS

The public health effects associated with lead are well established. An appropriate governmental function may be, for example, that of pressing the PrCB industry to make every effort to solve the technical problems resulting from the substitution of lead in its products. During routine maintenance, printed circuit boards are replaced periodically because they collect moisture over time. This offers the Department of Defense an opportunity to put policy into practice in a timely manner. In doing so, DoD will need to consider the impacts of environmentally unsafe practices occurring in PrCB fabrication as well as the technical challenges associated with the substitution of lead in PrCBs.

# Appendix F

# Sample Fabrication Sequence for a Standard Printed Circuit Board

The following sequence of steps helps convey the complexity of the process of fabricating a standard printed circuit board.

**Purchased Copper Clad Dielectric**

- Copper clad is measured in ounces, which converts to inches of thickness (1 oz is ~0.0014 in. thick).
- Clad may be on one or both sides of dielectric.
- Dielectric materials are determined by the design requirement. Common materials are glass-reinforced epoxies, glass-reinforced polyimides, polyimide films, and PTFE (Teflon).

**Preparation of the Copper Surface**

- Cleaning can be done mechanically or chemically.
- Photoresist is a coating of photosensitive material to be used to place the image of the inner-layer circuit pattern on the clad material.

---

NOTE: The figures and text in this appendix are drawn from educational materials available at http://www.rockwellcollins.com/about/additionalproducts/collinsprintedcircuits/making_circuit_boards/index.html and are used by permission from Rockwell Collins, Cedar Rapids, Iowa.

**Photo Process**

- *Expose* is the photo process of transferring the circuit image, which is on working tools, to the copper surface. The photo light causes the resist to harden (polymerize) and to be retained during the subsequent process steps.

- *Develop* chemically removes the remaining resist that was not exposed to the light source, as defined by the tooling.

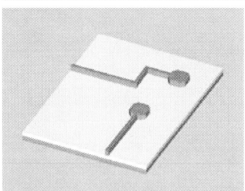

**Inner Circuitry**

- *Etch* is the chemical process that removes the unwanted copper in order to define the circuit image.

- The etch process will not attack the copper protected by the photoresist.

- Some amount under the resist may etch, thus causing the cross-section view of the circuitry to have an hourglass appearance.

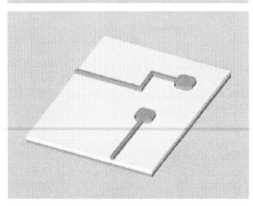

**Inner Circuitry**

- *Strip* is a chemical process that washes off the remaining photoresist leaving the circuit pattern in copper.

**Bond Package**

- *Layup* is the procedure by which a multilayer package is assembled. The circuit layers (laminates) are stacked over pins in plates, in numerical circuit order. Between each laminate is placed a predefined thickness of bonding material, usually a multiple of sheets.

### Bonding

- The *lamination* process is a temperature and pressure method, typically performed under vacuum.
- During lamination the resin in the bond material liquefies and then cures back to a solid state creating a homogenous package.
- Although different bonding and laminate materials may be interchanged, the best results are usually found when using common dielectrics that are compatible.

### Interconnection

- *Drilling* of holes using the design data for size and location is the start of the interconnect process. Drilling provides the mechanical opening that contacts or clears internal features. In the case of many internal layers (multilayer), the contact points are now the avenue of electrical communication.
- *Hole clean* is the process (usually chemical) that removes any internal debris from the internal connection. Depending on the materials and specifications, the walls can be simply cleaned (*desmear*) or aggressively attacked (*etchback*).

### Plated Connection

- Cleaning and activating solutions prepare the internal hole surfaces for copper plating.
- *Electroless copper* chemically plates an initial thin copper layer over the whole panel surface. This provides a surface that can now be electrically plated, which will provide a thicker amount of copper to build a tublet-type hole.
- *Strike plate* is the electroplate process that deposits additional amounts of copper in the holes and on the surface. (This process is optional.)

### Preparation for Outer Features

- This step is similar to the inner process of surface preparation, except that here the plan is to define the outer features. The clean and photoresist application steps are very similar.
- The coating process can be used with varying thicknesses of resists to allow for specific photo and plating process operations. The resist thickness is usually greater when multiple, subsequent plating steps are involved.

### Photo Process

- The steps in this process are similar to those for the inner process. The *expose* process, although similar, has the basic concept reversed. The photo tooling has been made to allow the polymerized areas to be where copper will be later removed.
- *Develop* will chemically remove the resist on the surface features, opening a pocket for additional plating and revealing the prior plated copper.

### Surface Finish

- *Pattern plate* means the plating of defined features. It is the process by which additional copper is electrically deposited in the developed cavities.
- This final copper plate will bring hole sizes and surface-feature thicknesses up to customer specifications.
- This copper-plated surface is then overplated, normally with a type of tin-lead; nickel or gold are also used as specified by design.

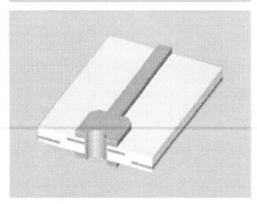

### Outer Circuitry

- The *etch* chemical process now removes all of the unwanted copper as supplied by the raw material vendor and plated on at electroless and strike (panel) operations.
- Similar to the inner etch, a surface cross-section view of the circuitries will show hourglassing.
- At this point in the build process the design is electrically completed.
- The only steps remaining are those of finishing the surface requirements, creating the mechanical size features, and the final inspecting.

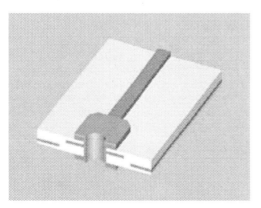

### Photo Expose and Strip Tin-Lead

- The tin-lead strip process involves cleaning the panel, laminating the photoresist on the panel, exposing and developing a pattern to selectively strip the tin-lead. This prepares the copper lines for soldermask to become the protective coating over the bare copper lines.

### Liquid Photo-Imageable Soldermask

- Soldermask is a protective layer of acrylic-epoxy (normally green) applied over bare copper lines. The process is known as solder mask over bare copper (SMOBC). The process for applying soldermask is clean, coat, expose, develop, and cure.

### Solderable Finishes

- The earliest final finish was *reflowed tin-lead,* in which tin-lead etch resist was melted using hot oil to form a solderable finish.

- Next came solder coating using a process called *hot air solder leveling (HASL),* in which boards are dipped in molten solder and withdrawn through forced air, which leveled the solder coating.

- Lately, immersion finishes have gained favor owing to the planar deposit they provide. Planar deposits help placement and soldering of tiny components.

### Identification

- A silkscreen legend is either a customer requirement for component identification, or it is completed by the printed circuit board supplier and used for board identification.

- Colors of ink are normally black, white, yellow, or red.